CONTENTS

The EARTH TRANSFORMED

B

The EARTH TRANSFORMED

An Introduction to Human Impacts on the Environment

ANDREW GOUDIE & HEATHER VILES

BLACKWELL *Publishers*

First published 1997
Reprinted 1998, 2000 and 2003

Blackwell Publishers Ltd
108 Cowley Road
Oxford OX4 1JF
UK

Blackwell Publishers Inc.
350 Main Street
Malden, Massachusetts 02148
USA

British Library Cataloguing in Publication Data

A CIP catalogue record for this book is available from the British Library.

Library of Congress Cataloging-in-Publication Data

Goudie, Andrew.
   The earth transformed: an introduction to human impacts on the environment/by Andrew Goudie and Heather Viles.
        p.   cm.
   Includes bibliographical references and index.
   ISBN 0–631–19464–9. —ISBN 0–631–19465–7 (pbk. : alk. paper)
   1. Nature—Effect of human beings on.   I. Viles, Heather A.
II. Title.
GF75G677   1997
304.2'8—DC20   96–26798
                                                       CIP

Typeset in 10 on 12 pt Galliard
Printed by Athenaeum Press Ltd., Gateshead, Tyne & Wear

This book is printed on acid-free paper

PART I

Introduction to the Developing Environmental Impact

1 EARLY DAYS

In this book we explore the many ways in which humans have transformed the face of the Earth. We start by placing these transformations into an historical context and seeing how they have changed through time.

Human life probably first appeared on Earth during the early part of the Ice Age, some 3 million years ago. The oldest human remains have been found in eastern and southern Africa. For a very long time the numbers of humans on the planet were small, and even as recently as 10,000 years ago the global population was probably only about one-thousandth of its size today. Also, for much of that time humans had only modest technology and limited capacity to harness energy. These factors combined to keep the impact of humans on the environment relatively small. Nonetheless, early humans were not totally powerless. Their stone, bone and wood tool technology developed through time, improving their efficiency as hunters. They may have caused marked changes in the numbers of some species of animals and in some cases even their extinction (see part II, section 13). No less important was the deliberate use of fire (see part II, section 2), a technological development that may have been acquired some 1.4 million years ago. Fire may have enabled even small human groups to change the pattern of vegetation over large areas.

2 DEVELOPING POPULATIONS

There are at least three interpretations of global population trends over the last 3 million years (Whitmore et al., 1990).

Plate I.1 The Olduvai Gorge in Tanzania is one of a group of sites in the Rift Valley of East Africa where some of the earliest remains of humans and their stone tools have been found. (A. S. Goudie)

Plate I.2   A grass fire in the high grasslands of Swaziland, southern Africa. Fire was one of the first ways in which humans transformed their environment and was probably used deliberately in Africa over a million years ago. (A. S. Goudie)

The first, described as the 'arithmetic-exponential' view, sees the history of global population as a two-stage phenomenon: the first stage is one of slow growth, while the second stage, related to the industrial revolution (see section 4 below), displays a staggering acceleration in growth rates. The second view, described as 'logarithmic-logistic', sees the last million or so years in terms of three revolutions – the tool, agricultural and industrial revolutions. In this view, humans have increased the **carrying capacity** of the Earth at least three times. There is also a third view, described as 'arithmetic-logistic', which sees the global population history over the last 12,000 years as a set of three cycles: the 'primary cycle', the 'medieval cycle' and the 'modernization cycle'. These three alternative models are presented graphically in figure I.1.

3   Agricultural Revolutions

Until the beginning of the **Holocene**, about 10,000 years ago, humans were primarily hunters and gatherers. After that time, in various parts of the world, increasing numbers of them started to keep animals and grow plants. **Domestication** caused **genetic** changes in plants and animals as people tried to breed more useful, better-tasting types. Domestication also meant that human populations could produce more reliable supplies of food from a much smaller area than hunter–gatherers (table I.1). This in turn created a more solid and secure foundation for cultural advance, and allowed a great increase in population density. This phase of development is often called the first agricultural revolution.

As the Holocene progressed, many other

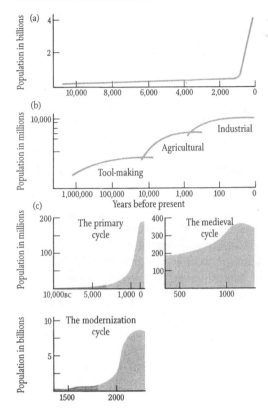

Figure I.1 Three interpretations of global population trends over the millennia: (a) the arithmetic-exponential; (b) the logarithmic-logistic; (c) the arithmetic-logistic
Source: Whitmore et al. (1990), figure 2.1.

technological developments occurred with increasing rapidity. All of them served to increase the power of humans to modify the surface of the Earth. One highly important development, with rapid and early effects on environment, was irrigation. This was introduced in the Nile Valley and Middle East over 5,000 years ago. At around the same time the plough was first used, disturbing the soil as never before. Animals were used increasingly to pull ploughs and carts, to lift water and to carry produce. Altogether the introduction of intensive cultivation and intensive

pastoralism (the use of land for keeping animals) had a profound effect on many environments in many parts of the world.

A further significant development in human cultural and technological life was the mining of ores and the smelting of metals, begun around 6,000 years ago. Metal artefacts gave humans greater power to alter the environment. The smelting process required large quantities of wood which caused local deforestation.

4   URBAN AND INDUSTRIAL REVOLUTIONS

The processes of urbanization and industrialization are two other fundamental developments that have major environmental implications. Even in ancient times, some cities evolved with considerable populations. Nineveh (the Assyrian capital) may have had a population of 700,000, Augustan Rome may have had a population of around 1 million, and Carthage (on the North African coast), at its fall in 146 BC, had 700,000 inhabitants. Such cities would have exercised a considerable influence on their environs, but this influence was never as extensive as that of cities in the last few centuries. The modern era, especially since the late seventeenth century, has witnessed the transformation of culture and technology through the development of major industries (table I.2). This 'industrial revolution', like the agricultural revolution, has reduced the space required to sustain each individual and has seen resources utilized more intensively.

Part of this industrial and economic transformation was the development of successful ocean-going ships in the sixteenth and seventeenth centuries. As a result, during this time countries in very different parts of the world became increasingly interconnected. Among other things, this gave humans the power to

Table I.1   Five stages of economic development

Economic stage	Dates and characteristics
Hunting–gathering and early agriculture	Domestication first fully established in south-western Asia around 7500 BCE; hunter–gatherers persisted in diminishing numbers until today. Hunter–gatherers generally manipulate the environment less than later cultures, and adapt closely to environmental conditions.
Riverine civilizations	Great irrigation-based economies lasting from c.4000 BC to 1st century AD in places such as the Nile Valley and Mesopotamia. Technology developed to attempt to free civilizations from some of the constraints of a dry season.
Agricultural empires	From 500 BC to around 1800 AD a number of city-dominated empires existed, often affecting large areas of the globe. Technology (e.g. terracing and selective breeding) developed to help overcome environmental barriers to increased production.
The Atlantic–industrial era	From c.1800 AD to today a belt of cities from Chicago to Beirut, and around the Asian shores to Tokyo, form an economic core area based primarily on fossil fuel use. Societies have increasingly divorced themselves from the natural environment, through air conditioning for example. These societies have also had major impacts on the environment.
The Pacific–global era	Since the 1960s there has been a shifting emphasis to the Pacific Basin as the primary focus of the global economy, accompanied by globalization of communications and the growth of multinational corporations.

Source: Adapted from Simmons (1993), pp. 2–3.

Plate I.3   A simple irrigation system in use in the drier portions of Pakistan. Such irrigation was probably introduced in the Old World drylands around 5,000–6,000 years ago. (A. S. Goudie)

introduce plants and animals to parts of the world where they had not previously been. The steam engine was invented in the late eighteenth century and the internal combustion engine in the late nineteenth century: both these innovations massively increased human need for and access to energy, and lessened dependence on animals, wind and water.

5   THE MODERN SCENE

Modern science and modern medicine have compounded the effects of the urban and industrial revolutions, leading to accelerating population increase even in nonindustrial societies. Urbanization has gone on speedily, and it is now recognized that large cities have their own environmental problems, and produce a multitude of environmental effects. If present trends continue, many cities in the less developed countries will become unimaginably large and crowded. For instance, it is projected that by the year 2000 Mexico City will have more than 30 million people – roughly three times the present population of the New York metropolitan area. Calcutta, Greater Bombay, Greater Cairo, Jakarta and Seoul are each expected to be in the 15–20 million range by that time. In all, around 400 cities will have passed the million mark by the end of the twentieth century, and UN estimates indicate that by then over 3,000 million people will live in cities, compared with around 1,400 million people in 1970.

Modern science, technology and industry have also been applied to agriculture. In recent decades some spectacular progress has been made. Examples include

Plate I.4 A limestone pavement developed on the Carboniferous limestone of north-west England. Although they were formed in glacial times by glacial abrasion, they may be exposed at the surface today because of soil erosion produced by forest clearance since the Mesolithic. (A. S. Goudie)

the use of fertilizers and the selective breeding of plants and animals. **Biotechnology** has, however, immense potential to cause environmental change (see part II, section 14).

We can recognize certain trends in human manipulation of the environment during the modern era. First, the number of ways in which humans are affecting the environment is growing rapidly. For example, nearly all the powerful pesticides post-date the Second World War. The same applies to the increasing construction of nuclear reactors, to the use of jet aircraft and to many aspects of biotechnology. Secondly, environmental issues that once affected only particular local areas have become regional or even global problems.

An instance of this is the appearance of substances such as DDT (a major pesticide), lead and sulphates at the North and South Poles, far removed from the industrial societies that produced them. Thirdly, the complexity, magnitude and frequency of impacts are probably increasing. For instance, a massive modern dam like that at Aswan in Egypt has a very different impact from a small Roman dam. Finally, a general increase in **per capita** consumption and environmental impact is compounding the effects of rapidly expanding populations. Energy resources are being developed at an ever-increasing rate, giving humans enormous power to transform the environment. One measure of this is world commercial energy consumption, which

Table 1.2 Energy, technology and environmental impact time line

Time zone	Global population	Daily energy use per person (kcals)	Energy source	Technological discoveries	Environmental impacts
1 million to 5000 years BC	< 10 million	2,000–5,000	Food, human muscle	Tool production, fire	Local and short-term; animal kills and vegetation change
5000 BC to AD 1800	10 million –1 billion	12,000– 26,000	Animals, agricultural crops, wind, water, coal	Cultivation, building, transport, irrigation	Local and longer-term; natural vegetation removal, soil erosion, urban air pollution
AD 1800 to 1950	1 billion– 4 billion	50,000	Fossil fuels, electricity, steam	Industry	Local, regional and permanent; major landscape changes, air and water pollution common
1950 to present	>4 billion	300,000	Internal combustion engine, electricity, nuclear, fossil fuels	Industry, cultural globalization	Local, regional, global; permanent and perhaps irreversible, acid rain, global warming

Plate I.5 The power of humans to transform the land's surface in the modern era is illustrated by the size of the giant open-cast uranium mine at Rössing, Namibia. Modern technology allows humans to harness energy resources as never before. (A. S. Goudie)

trebled in size between the 1950s and 1980.

The importance of the harnessing of energy can be clearly seen in the context of world agriculture. At the beginning of the twentieth century, more or less throughout the world, farmers relied upon domestic animals to provide both pulling power and fertilizer. They were largely self-sufficient in energy. However, in many areas the situation has now changed. Fossil fuels are extensively used to carry out such tasks as pumping (or, in many cases, mining) water, propelling tractors and manufacturing synthetic fertilizers (which in many cases cause pollution). The world's tractor fleet has quadrupled since 1950 and as much as two-thirds of the world's cropland is being ploughed and compacted by increasingly large tractors.

Above all, as a result of the huge expansion of environmental transformation it is now possible to talk about *global* environmental change. There are two aspects of this (Turner, Kasperson et al., 1990): 'systemic' global change and 'cumulative' global change. Systemic global change refers to changes operating at the global scale and includes, for example, global changes in climate brought about by atmospheric pollution, e.g. the **greenhouse effect** (see part III). Cumulative global change refers to the snowballing effect of local changes, which add up to produce change on a worldwide scale, or change which affects a significant part of a specific global resource, e.g. **acid rain** or soil erosion (see parts III and V). The two types of change are closely linked. For example, the burning of vegetation can lead to systemic global change through

Table I.3 Systemic and cumulative global environmental changes

Type of change	Characteristic	Examples
Systemic	Direct impact on globally functioning system	(a) Industrial and land-use emissions of 'greenhouse' gases (b) Industrial and consumer emissions of ozone-depleting gases (c) Land cover changes in albedo
Cumulative	Impact through worldwide distribution of change	(a) Groundwater pollution and depletion (b) Species depletion/genetic alteration (biodiversity)
	Impact through magnitude of change (share of global resource)	(a) Deforestation (b) Industrial toxic pollutants (c) Soil depletion on prime agricultural lands

Source: Turner, Clark et al. (1990), table 1.

processes such as carbon dioxide release and **albedo** modification, and to cumulative global change through its impact on soil erosion and **biodiversity** (table I.3).

Figure I.2 shows how the human impact on six 'component indicators of the biosphere' has increased over time. This graph is based on work by Kates et al. (1990). For each component indicator they defined the total net change clearly induced by humans to be 0 per cent for 10,000 years ago (before the present–BP) and 100 per cent for 1985. They then estimated the dates by which each component had reached successive quartiles (that is, 25, 50 and 75 per cent) of its total change at 1985. They believe that about half of the components have changed more in the single generation since 1950 than in the whole of human history before that date.

Human activities are now causing environmental transformation on the local, regional, continental and planetary scales. The following examples both give an indication of what is currently being achieved and provide a sample of some of the issues we cover in this book.

Large areas of **temperate** forest have been cleared in the past few centuries. Now farmers and foresters are removing forests from the humid tropics at rates of around 11 million hectares (ha) per year. This is exposing soils to intense and erosive rainfall and increasing rates of **sediment yield** by an average of six times. The world's rivers are being dammed by around 800 major new structures each year, transforming downstream **sediment loads**. Huge reservoirs held behind dams as high as 300 metres are generating seismic hazards and catastrophic slope failures. Some of the world's largest lakes, most notably the Aral Sea in the former Soviet Union, are becoming **desiccated** because the water is being taken for irrigation use and transferred to other water basins at a near-continental scale. Fluids, both water and **hydrocarbons** (e.g. oil and gas) are being withdrawn from beneath cities and farmlands, leading to subsidence of up to 8–9 metres. Recreational vehicles and trampling

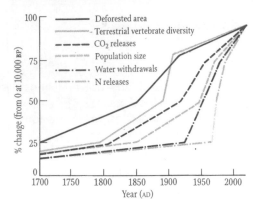

Figure I.2 Percentage change (from assumed zero human impact at 10,000 BP) of selected human impacts on the environment

feet are damaging many popular tourist areas. Development on **tundra** areas is disturbing the thermal equilibrium of **permafrost**, leading to more and more instances of **thermokarst**. Coastlines are being 'protected' and 'reclaimed' by the use of large engineering structures, often without due thought for the possible consequences. We are pumping at least 500 million tonnes of dissolved material into rivers and oceans around the world each year. We are acidifying **precipitation** to the extent that some of it has the **pH** of vinegar or stomach fluid, thereby altering rates of mineral release and rock weathering.

These human impacts are having great direct and indirect effects on vegetation: table I.4 shows the amounts of vegetation (in terms of **net primary production**) used, dominated or lost by humans.

We shall return to these and other issues in subsequent sections. In this book we have chosen to focus on specific environmental issues as they affect the **biosphere** (part II), atmosphere (part III), surface waters (part IV), land surface (part V), and oceans, seas and coasts (part VI). However, you will notice through all of these sections that a range of important human activities play key roles and can have a range of different impacts on many sectors of the environment.

Even in the modern world economy, hunting and gathering activities still have an important effect on the environment, largely through the biological impacts of fishing and the shooting of game. These activities are becoming increasingly large-scale and mechanized. Agriculture, **aquaculture** and other forms of food production now occupy vast areas of the Earth's surface and have a wide variety of environmental effects, including soil erosion, nutrient depletion, changes in species diversity and genetic changes to crops and animals. Forestry and quarrying, as extractive industries, are creating whole new landscapes and releasing large amounts of sediment in parts of the globe ranging from the humid tropics to the Arctic. Heavy industries (such as oil refining and chemical manufacture), power generation plants (from coal-fired to nuclear), and light and high-technology industries have many different environmental impacts and contribute to pollution of land, water and air on the local and regional scales.

Transport and urbanization have, perhaps, some of the most dramatic local impacts on the environment. They create whole new landscapes dominated by concrete, add to pollution, and affect plant and animal distributions and the circulation and distribution of water. Tourism, which is now a booming global industry, also has considerable impacts on the environment. In recent years there has been much interest in the notion of 'ecotourism', or tourism which attempts to minimize environmental damage.

One of the consequences of all these different human interactions with the environment is the production of waste. This itself has had major environmental effects. There are problems of waste disposal and

Table I.4 Terrestrial net primary production of vegetation used, dominated or lost through human activities

Category	Amount (Pg per year)[a]
NPP used:	
consumed by humans	0.8
consumed by domestic animals	2.2
wood used by humans	2.4
Total	5.2 (4% of total global NPP)
NPP dominated:	
croplands	15
converted pastures	10
tree plantations	2.6
human-occupied lands	0.4
consumed from little-managed areas	3
land-clearing	10
Total	41 (31% of total global NPP)
NPP lost to human activities:	
decreased NPP of cropland	10
desertification	4.5
human-occupied areas	2.6
Total	17 (8% of total global NPP)
Total NPP dominated and lost	58 (39% of total global NPP)

[a] 1 Pg or Petagramme = 1×10^{15} g.
Source: Vitousek (1994).

waste management. Big issues like nuclear waste disposal have potentially long-term environmental implications. So do less contentious matters, such as disposing of domestic and industrial waste on **landfill** sites.

Human societies do not always run smoothly. War, civil strife and smaller-scale disruptions such as vandalism and crime have their own environmental conse- quences. Indeed, some wars are partly motivated by disputes over environmental resources, for example over water supplies. Recent conflicts in the Arabian Gulf, Bosnia and Afghanistan have had both short-term and long-term environmental consequences, including pollution and soil erosion. In the 1960s and 1970s the Vietnam War had widely publicized effects on the **mangrove** vegetation of the Mekong Delta. The use of **defoliant** chemicals there has had long-term impacts on biodiversity from which the envir onment is only just recovering. Even without war, political systems can impose additional stress on the environment. The apartheid system in pre-1994 South Africa, for example, forcibly distributed popu- lation and wealth in a highly unfair way, leading to huge environmental pressures on **marginal land**. The planned socialist economies of the former Soviet Union and many East European states appear now to have had particularly damaging environ- mental impacts. And capitalist enterprise, which now dominates the global economy, has often had a tendency to plunder and despoil the environment.

These many negative environmental impacts have generated in response a long-term, and growing, focus on conservation and improving human management of the environment. Conservation and management themselves have environmental impacts, as in the creation of nature reserves; there may also be less desirable impacts where management schemes go wrong. The ideas of **sustainable development** are the most recent attempt to combine resource exploitation with conservation and a concern for the environmental future. As our scientific understanding of how the environment works has advanced, we have gained a better view of how serious our human impacts can be. On the other hand, we have also learnt that there is much reason for hope. The environmental future is not all doom and gloom, as we stress in part VII of this book.

6   UNDERSTANDING ENVIRONMENTAL TRANSFORMATIONS

We have already shown in this chapter that human impacts on environmental processes have had a long and complex history, and now take on many complex and inter-linked forms. The environment itself is also not a static, simple entity, but has a complicated history of its own. We now realize that the environment changes naturally, over a range of different time-scales, as a response to a number of natural 'forcing factors', such as the varying position of the Earth within its orbit around the sun. On shorter time-spans, we know that the environment can work in abrupt and challenging ways, producing what are called 'natural hazards' such as volcanic eruptions, earthquakes, floods and hurricanes. So, putting together human and natural factors influencing the environment to

explain any single environmental transformation can be a hugely difficult task. It is important to realize that there is still a lot of scientific uncertainty and debate over the causes and consequences of many of the environmental issues we look at in this book.

Understanding the role of human activities in environmental transformations is not a completely hopeless task, however. There are several useful concepts which we can adopt to help us untangle what is going on. First, it is useful to think of the environment (of which, of course, we are a part) as being a series of interlocked systems. These systems are affected by a whole series of stresses (which can be human or natural in origin). The stresses produce some changes in the system, or responses: these are what we see as environmental transformations or environmental issues. Because the systems are interlocked, stresses on one system may produce linked effects on other systems. Some of the systems are more able to resist stresses than others, and so some can be stressed greatly before they show any response. Others are more sensitive to stresses.

As an example, to clarify the ideas presented above, we could look at a **drainage basin** (or **watershed**, as it is known in the USA). Drainage basins are primarily **hydrological** systems, with interlinked vegetation communities. Cutting down trees (a stress) will produce a range of responses: soil erosion, increased flooding and changes in the way water is distributed (hydrological pathways). The severity of these outcomes will depend on the climate and topography of the area. Normally, a mixture of natural and human-induced stresses will affect the environment together, complicating the picture. One way of understanding such multi-causal situations is to identify different types of stresses or causal factors. A useful framework, which has been used in various ways in the following sections of this book, is to split causal factors into three types: that is, 'predisposing', 'inciting' and 'contributing' factors. *Predisposing* factors are those features of the natural or human environment which make a system vulnerable to stress; *inciting* factors are those stresses that trigger off a change in the system; and *contributing* factors are the whole range of additional stresses which make the response more noticeable and acute. Let us apply this framework to the case of a drainage basin. The *predisposing* factors which may make it vulnerable to change following tree-cutting would be the topography and climate, and perhaps also past forest management practices. The *inciting* factor would be the tree-cutting itself. The *contributing* factors could be the health of the trees; the season when the trees were felled; the weather at the time; and, over a longer time-span, what vegetation grows in place of the trees.

The concepts of stresses, responses and different types of causal factors are very useful in trying to understand how humans are influencing their environment. Such understanding is vital in any attempts to solve or manage the resultant environmental problems. However, to arrive at solutions it is also necessary to have a deeper understanding of the human societies involved in such environmental issues, as many of our subsequent examples illustrate. For example, just knowing how tree-cutting can produce soil erosion and hydrological changes within a drainage basin does not mean that we can solve the problem. We need also to know why people are cutting down the trees. Before we can effect any great changes, we need more understanding of the economic conditions, technological capability, cultural organization and political systems of the people involved.

FURTHER READING

Freedman, B., 1995, *Environmental Ecology*, 2nd edn. San Diego: Academic Press.
An enormously impressive and wide-ranging study with a strong ecological emphasis.

Mannion, A. M., 1995, *Agriculture and Environmental Change*. London: Wiley.
A new and comprehensive study of the important role that agriculture plays in land transformation.

Meyer, W. B., 1996, *Human Impact on the Earth*. Cambridge: Cambridge University Press.
A good point of entry to the literature that brims over with thought-provoking epigrams.

Middleton, N. J., 1995, *The Global Casino*. London: Edward Arnold.
An introductory text, by a geographer, which is well illustrated and clearly written.

Ponting, C., 1991, *A Green History of the World*. London: Penguin.
An engaging and informative treatment of how humans have transformed the earth through time.

Simmons, I. G., 1996, *Changing the Face of the Earth: Culture, Environment and History*, 2nd edn. Oxford: Blackwell.
A characteristically amusing and perceptive review of many facets of the role of humans in transforming the earth, from an essentially historical perspective.

KEY TERMS AND CONCEPTS

agricultural revolution
biosphere
global environmental change
Holocene
hunter–gatherer

industrial revolution
predisposing, inciting and contributing
  factors
stresses and responses
systems

POINTS FOR REVIEW

How much environmental change was achieved in prehistoric times and how much in the last three centuries?

To what extent are environmental changes the result of both natural and human-induced stresses?

What do you understand by the phrases 'global environmental change' and 'sustainability'?

PART II

The Biosphere

The Biosphere

1   INTRODUCTION

In this part of the book we look at some of the main ways in which humans have modified the biosphere, and the consequences of these impacts.

Humans have changed the biosphere in very many ways, with wide-ranging and long-lasting effects. As soon as people discovered how to use fire, at a very early stage in human development, they obtained tremendous power to modify the vegetation cover of the Earth's surface. Also, during the Stone Age, humans gradually developed the technology to enable them to become ever more effective hunters. Early people may have contributed to the extinction of some of the world's great **mammals**. Since the **Mesolithic**, as pastoralism and agriculture have become widespread, modification of **habitat** has continued rapidly. Humans also gained the ability to manipulate the genetic composition of plants and animals – a major part of the process generally called domestication. This has been one of the most direct ways that humans have changed the biosphere.

As the human population of the Earth has expanded in numbers and spread to more and more parts of the globe, ever more environments have been modified. These include tundra areas, deserts, forests and **wetlands**. The total area of surviving 'natural' habitat has steadily diminished and **wilderness** areas are now relatively few. Figure II.1 shows an attempt to mark out the areas of the planet that can still be defined as wilderness. However, no part of the Earth's surface can be considered entirely free from the imprint of human activities. Air pollution and climatic changes caused by human action are evident even at the poles. As it has become easier for humans to move from one place to another, so plants and animals have been introduced to many new areas. Sometimes the numbers of these newcomer species have exploded, damaging the community structure of existing plants and animals.

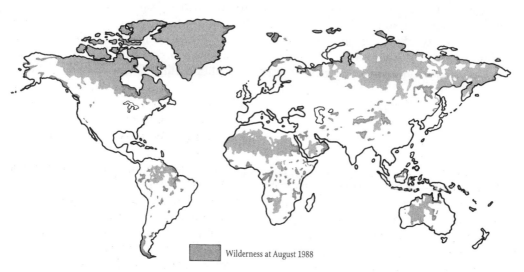

Wilderness at August 1988

Figure II.1   Global wilderness remaining in the 1980s
Source: McCloskey and Spalding (1989).

Table II.1   Biomass burning in the tropical regions

Region	Forest	Savanna	Fuel wood	Agricultural waste	Regional total	Regional total
	(Tg dm/yr)	(Tg dm/yr)	(Tg dm/yr)	(Tg dm/yr)	(Tg dm/yr)	(Tg C/yr)
America	590	770	170	200	1,730	780
Africa	390	2,430	240	160	3,210	1,450
Asia	280	70	850	990	2,190	980
Oceania	–	420	8	17	450	200
Total tropics	1,260	3,690	1,260	1,360	7,580	3,410

Tg dm/yr = teragrammes of dry matter per year.
Tg C/yr = teragrammes of carbon per year.
Source: Andreae (1991), table 1.3.

2   FIRE

Fire is one of the earliest means that humans used to modify the natural environment. It is also one of the most powerful. Fires do, of course, occur naturally, and have done so during the entire history of the Earth. For example, they are caused by volcanic eruptions, by **spontaneous combustion** of organic materials, by sparks from falling boulders and, above all, by lightning, which on average strikes the land surface of the globe 100,000 times each day. However, in some environments the great majority of fires are now caused by humans, either deliberately or accidentally.

There are many good reasons why humans, from our early Stone Age ancestors onwards, have found fire useful:

- to clear forest for agriculture;
- to improve the quality of grazing for game or domestic animals;
- to deprive game of cover or to drive them from cover;
- to kill or drive away **predatory** animals, insects and other pests;
- to repel or attack human enemies;
- to make travel quicker and easier;
- to provide light and heat;
- to enable them to cook;

- to transmit messages, by smoke signs;
- to break up stone for making tools or pottery, smelting ores, and hardening spears or arrowheads;
- to make charcoal;
- to protect settlements or camps from larger fires by controlled burning;
- to provide spectacle and comfort.

Fire has been central to the life of many groups of hunter–gatherers, pastoralists and farmers (including **shifting cultivators** in the tropics). It was much used by peoples as different from one another as the Aboriginals of Australia, the cattle-keepers of Africa, the original inhabitants of Tierra del Fuego ('the land of fire') in the far south of South America and the Polynesian inhabitants of New Zealand. It is still much used, especially in the tropics, and above all in Africa. **Biomass** burning appears to be especially significant in the tropical environments of Africa in comparison with other tropical areas (table II.1). The main reason for this is the great extent of **savanna** which is subjected to regular burning. As much as 75 per cent of African savanna areas may be burned each year (Andreae, 1991). This is probably an ancient phenomenon in the African landscape which occurred long before people arrived on the scene. Nevertheless, humans have greatly increased the role of fire in the continent, where they may have used it for over 1.4 million years (Gowlett et al. 1981).

Naturally occurring fires break out with varying frequency in different global environments. Over a century may pass between one fire and the next in tundra environments and **ecosystems** dominated by the spruce tree. In areas of savanna and Mediterranean shrubland, on the other hand, the interval may be only five to fifteen years, and in **semi-arid** grasslands less than five years.

Fires can extend over huge areas. In 1963 in Parana, Brazil, no less than 2 million hectares of forest were consumed in just three weeks, while the fire of 1987 in China and the neighbouring Soviet Union destroyed around 5 million hectares over the same length of time.

Fires can also cause some very high ground surface temperatures, up to 800°C or higher. The temperature reached depends very much on the size, duration and intensity of the fire. Some fires are relatively quick and cool-burning, and only destroy ground vegetation. Other fires, such as 'crown fires', affect whole forests up to the level of tree crowns and generate very high temperatures. In general, forest fires are hotter than grassland fires. It is significant for forest management that where fires occur very often they do not attain the highest temperatures, because there is not enough flammable material to keep them going. However, humans often deliberately prevent fires as part of normal policy in forest areas. When this is done, large quantities of flammable materials accumulate, so that when a fire does break out it is of the hot, crown type that can be ecologically disastrous. There is now much debate, therefore, about the wisdom of suppressing the fires that in many forests would occur quite regularly under so-called 'natural' conditions.

Recent studies have indicated that rigid policies of protecting habitats against fire have often had undesirable results. Consequently, many foresters now stress the need for 'prescribed burning' or 'environmental restoration burning'. For example, in the coniferous forests of the middle and upper levels of the Sierra Nevada mountains of California, protection from fire since 1890 has made the stands denser, shadier and less park-like, and sequoia seedlings have decreased in number as a result. Likewise, at lower levels the character of the semi-arid shrubland, called **chaparral**, has changed. The vegetation has become denser, the amount of flammable material has increased, and fire-sensitive species have encroached. The vegetation has become less diverse, with older trees predominating instead of a mosaic of trees at different stages of growth. In the Kruger National Park, in South Africa, fires have become less frequent since the game reserve was established, when local hunters and farmers were moved out. As a result, bush has encroached on areas that were formerly grassland, and the carrying capacity for grazing animals has declined. Controlled burning has been reinstituted as a necessary element of game management.

Fire has many positive ecological consequences. Fire may assist in seed germination. For example, many investigators have reported the abundant germination of dormant seeds on recently burned chaparral in areas like California with a **Mediterranean climate**, and it seems that some seeds of chaparral species require **scarification** by fire to germinate effectively. Fire alters **seedbeds**, and even those seeds not requiring scarification may germinate better after a fire because fire removes competing seeds, litter and some substances in the soils which are toxic to plants. If substantial amounts of litter and **humus** are removed, large areas of rich ash, bare soil or thin humus may be created. Some trees, such as the Douglas fir and the giant sequoia, benefit from such seedbeds. Fire sometimes triggers the release of seeds from cones (as with Jack pine, *Pinus banksiana*) and seems to stimulate the vegetative reproduction of many woody and herbaceous species. Fire can control forest insects, parasites and fungi – a process termed **sanitization** – and seems to stimulate the flowering and fruiting of many shrubs and herbs. It also appears to modify the physiochemical environment of plants, with mineral elements being released both as ash and through faster decomposition of organic layers. Above all, areas subject to fire often show greater species diversity, which tends to favour the stability of the habitat over the long term.

Fire is also crucial to an understanding of some major **biome** types, and many **biota** have become adapted to it. For example, many savanna trees are fire-resistant. The same applies to the shrub vegetation (*maquis*) of the Mediterranean lands, which contains certain species (e.g. *Quercus ilex* and *Quercus coccifera*) which thrive after burning by sending up a series of suckers from ground level. Mid-latitude grasslands (e.g. the prairies of North America) were once thought to have developed in response to drought conditions during much of the year. Now, however, some have argued that this is not necessarily the case and that in the absence of fire, trees could become dominant. The following reasons are given to support this suggestion:

- planted groves and protected trees seem able to flourish;
- some woodland species, notably junipers, are remarkably drought-resistant;
- trees grow along escarpments and in deep valleys, where moisture is concentrated at **seeps** and in shaded areas,

and where fire is least effective: the effects of fire are greatest on flat plains where there are high wind speeds and no interruptions to the course of the fire;
- where fires have been restricted, woodland has spread into grassland.

Fire rapidly alters the amount, form and distribution of plant nutrients in ecosystems, and has been used deliberately to change the properties of the soil. Both the release of nutrients by fire and the value of ash have long been recognized, notably by those involved in shifting cultivation based on **slash-and-burn** techniques. However, once land has been cultivated, the loss of nutrients by **leaching** and erosion is very rapid. This is why the shifting cultivators have to move on to new plots after only a few years. Fire quickly releases some nutrients from the soil in a form that plants can absorb. The normal biological decay of plant remains releases nutrients more slowly. The amounts of phosphorus (P), magnesium (Mg), potassium (K) and calcium (Ca) released by burning forest and **scrub** vegetation are high in relation to both the total and the available quantities of these elements in soils.

In forests, burning often causes the pH value of the soil to rise by three units or more, creating alkaline conditions where formerly there was acidity. Burning also leads to some direct loss of nutrients from the soil: by **volatilization** and by causing ash to rise up into the air, or by loss of ash to water erosion or wind **deflation**. Where fire removes trees, soil temperatures increase because of the absence of shade, so that humus is often lost at a faster rate than it is formed.

Concern is now being expressed about the role of biomass burning in altering atmospheric chemistry and contributing to the greenhouse effect by adding carbon dioxide (CO_2) to the atmosphere (Levine, 1991). About 40 per cent of the world's annual production of CO_2 may result from the destruction of biomass by fire. Fires also produce emissions of smoke and nitric oxide.

FURTHER READING

Crutzen, P. J. and Goldammer, J. G., 1993, *Fire in the Environment*. Chichester: Wiley.
This book considers some of the potential global effects of fires, including effects on atmospheric chemistry.

Kozlowski, T. T. and Ahlgren, C. C. (eds), 1974, *Fire and Ecosystems*. New York: Academic Press.
Although relatively old, this provides a very useful picture of the effects of fire on fauna and flora.

Pyne, S. J., 1982, *Fire in America: A Cultural History of Wildland and Rural Fire*. Princeton: Princeton University Press.
A massive and scholarly survey of how fires have been fundamental to understanding much of the vegetation of America.

Plate II.1   Forest burnt in the Yellowstone fires of 1988. (EPL/Rob Franklin)

The Yellowstone fires of 1988

In the summer of 1988 widespread fires ravaged the Yellowstone National Park in the American West. Forest fires began in June and did not die out completely until the onset of winter in November. Somewhere between 290,000 and 570,000 hectares burned in by far the worst fire since Yellowstone was established as the world's first national park in the 1870s.

Was this inferno the result of a policy of fire suppression? Without such a policy the forest would burn at intervals of 10 or 20 years because of lightning strikes. Could it be that the suppression of fires over long periods of, say, 100 years or more, allegedly to protect and preserve the forest, led to the build-up of abnormal amounts of combustible fuel in the form of trees and shrubs in the **understory**? Should a programme of prescribed burning be carried out to reduce the amount of available fuel?

Fire suppression policies at Yellowstone did indeed lead to a critical build-up in flammable material. However, other factors must also be examined in explaining the severity of the fire. One of these was the fact that the last comparable fire had been in the 1700s, so that the Yellowstone forests had had nearly 300 years in which to become increasingly flammable. In other words, because of the way vegetation develops through time (a process called **succession**) very large fires may occur every 200–300 years as part of the natural order of things (figure II.2). Another crucial factor was that weather conditions in the summer of 1988 were abnormally dry, bringing a great danger of fire.

Romme and Despain (1989, p. 28) remark in conclusion to their study of the Yellowstone fires:

> It seems that unusually dry, hot and windy weather conditions in July and August of 1988 coincided with multiple ignitions in a forest that was at its most flammable stage of succession. Yet it is unlikely that past suppression efforts were a major factor in exacerbating the Yellowstone fire. If fires occur naturally at intervals ranging from 200 to 400 years, then 30 or 40 years of effective suppression is simply not enough for excessive quantities of fuel to build up. Major attempts at suppression in Yellowstone forests may have merely delayed the inevitable.

Further reading

Romme, W. H. and Despain, D. G., 1989, The Yellowstone fires. *Scientific American* 261, 21–9.

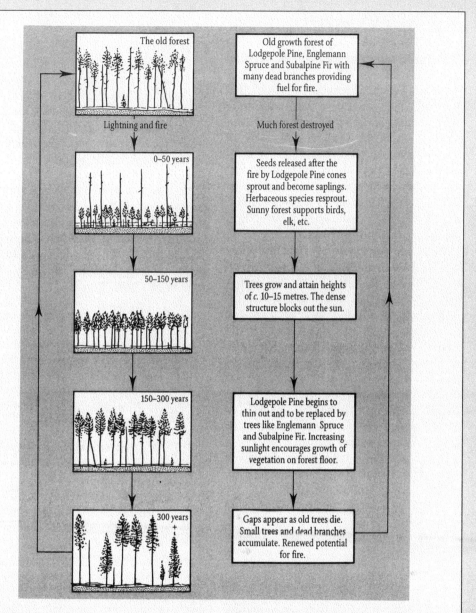

Figure II.2 Ecological succession in response to fire in Yellowstone National Park, USA

Source: After Romme and Despain (1989), pp. 24–5.

3   DESERTIFICATION

The term **desertification** was first used by the French forester Aubréville in 1949, but he never formally defined it. Since then over 100 definitions have been published. The United Nations Environment Programme (UNEP) has recently defined desertification as 'land degradation in **arid**, semi-arid and dry sub-humid areas resulting mainly from adverse human impacts' (Tolba and El-Kholy, 1992, p. 134). Others, however, suggest that climatic change may also play an important role.

There are fundamental problems relating to how extensive the problem of desertification is, how quickly it is taking place, and what the main causes are. UNEP (Tolba and El-Kholy, 1992, p. 134) has no doubts about the significance of the problem: 'Desertification is the main environmental problem of arid lands, which occupy more than 40 per cent of the total global land area. At present, desertification threatens about 3.6 billion hectares – 70% of potentially productive drylands, or nearly one-quarter of the total land area of the world. These figures exclude natural hyper-arid deserts. About one sixth of the world's population is affected.'

UNEP recognizes the following series of symptoms of desertification that relate to a fall in the biological and economic productivity and therefore value of a piece of land:

- reduction of crop yields (or complete failure of crops) in irrigated or rain-fed farmland;
- reduction of biomass produced by **rangeland** and consequent depletion of feed material available to livestock;
- reduction of available wood biomass, and consequent increase in the distances travelled to obtain fuelwood;
- reduction of available water due to decreases in river flow or **groundwater** resources;
- encroachment of sand bodies (dunes, sheets) that may overwhelm productive land, settlements or infrastructures;
- social disruption due to deterioration of life support systems, and the associated need for outside help (relief aid) or for havens elsewhere, producing 'environmental refugees'.

It is, however, by no means clear how extensive desertification is or how fast it is proceeding. In a recent book called *Desertification: Exploding the Myth*, Thomas and Middleton (1994) have discussed UNEP's views on the amount of land that is desertified. They state:

> The bases for such data are at best inaccurate and at worst centered on nothing better than guesswork . . . The advancing desert concept may have been useful as a publicity tool but it is not one that represents the real nature of desertification processes. (Thomas and Middleton, 1994, p. 160)

There are indeed relatively few reliable studies of the rate of desert advance or encroachment. Lamprey (1975) attempted to measure the shift of vegetation zones in the Sudan and concluded that the Sahara had advanced by 90–100 km between 1958 and 1975, an average rate of about 5.5 km per year. However, on the basis of data amassed by remote sensors and ground observation, Helldén (1984) found little evidence that this had in fact happened. One problem is that biomass production may vary very substantially from year to year. This has been revealed by satellite observations of green biomass production levels on the southern side of the Sahara.

The way in which desert-like conditions spread is also the subject of some controversy. Contrary to popular rumour, this

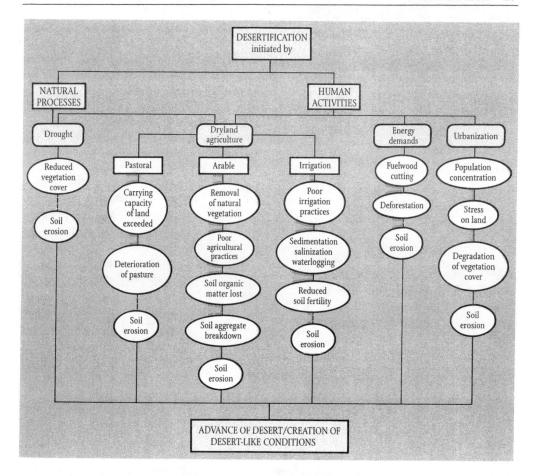

Figure II.3 The causes and development of desertification
Source: Kemp (1994), figure 3.12.

does not happen over a broad front, like a wave overwhelming a beach. Rather, it is like a 'rash', tending to appear in local patches around settlements. This distinction is important because it influences perceptions of how to tackle the problem.

Another point made by Thomas and Middleton (1994, p. 160) is that drylands may be less fragile than is often thought.

We should bear this controversy in mind as we consider some of the undoubted pressures that are being placed on arid environments (figure II.3). It is generally agreed that the massive increase in human population numbers during the twentieth century has been of fundamental importance. This demographic explosion has had four key consequences for dryland areas: overcultivation, overgrazing, deforestation and **salinization** of irrigation systems.

There are two aspects of overcultivation: more intensive use of land already under cultivation, and the introduction of agriculture into areas where conditions are not suitable to it, primarily because of their aridity or because their soils are fragile and infertile. Crops are now grown in areas of the Sahel of West Africa where annual rainfall is as low as 250 mm, and in parts of the Near East and North Africa which

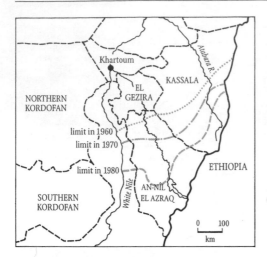

Figure II.4   The expanding wood and charcoal exploitation zone south of Khartoum, Sudan
Source: From Johnson and Lewis (1995), figure 6.2.

have only 150 mm of rainfall a year. Some of these areas have **friable** soils, developed on late **Pleistocene** dune fields. This makes them highly prone to water erosion and **wind reactivation**.

Overgrazing is related to overcultivation, for in many areas increasing numbers of humans require increasing numbers of domestic animals. Larger free-ranging stock herds reduce the amount of pastureland available and mean that the pastures that remain have to support even more animals. The carrying capacity of the land may then be exceeded. There may also be conflicts between pastoralists and cultivators. As the frontier of cultivation is pushed outwards into ever more marginal areas, it encroaches more and more on the grazing lands of the pastoralists. In this way, nomadic pastoralists, many of whom had developed sophisticated ways of keeping marginal areas in productive use, have often seen their traditional systems disintegrate. This has disrupted the equilibrium between people and land. For example, the nomads'

seasonal or annual migrations may have been restricted by deliberate policies of sedentarization (making people settle in one place) imposed by central governments. The same restrictive effect results from the establishment of national boundaries where none previously existed.

Another cause of overgrazing has been the installation of boreholes and the digging out of waterholes. These have made more water available for domestic animals, which thus rapidly increase in number. This in turn leads to overgrazing. Vegetation, in effect, replaces water as the main factor limiting stock numbers.

The third human cause of desertification is deforestation and the removal of woody material. Many people depend on wood for domestic uses (cooking, heating, brick manufacture, etc.), and the collection of wood for charcoal and firewood is an especially serious problem in the vicinity of urban centres. This is illustrated for Khartoum in Sudan in figure II.4.

The fourth prime cause of desertification is salinization. This kills plants, destroys the soil structure and reduces plant growth. Salinization often occurs where irrigation is introduced without making proper provision for drainage. It can, however, also be an unwanted consequence of vegetation clearance. The removal of plants reduces the amount of moisture lost from the soil as a result of interception of rainfall by leaves and **evapotranspiration**. As a result, groundwater levels rise and saline water is allowed to seep into low-lying areas like valley bottoms. This is a serious cause of salinization both in the prairies of North America and in the wheat belt of Western Australia. It is so important that we treat it in a separate section (part V, section 5 below).

As we saw earlier, some observers have suggested that a natural deterioration in the climate may contribute to the damage done to dryland and the spread of desert-like conditions. When we examine rainfall data for recent decades, we see that for some arid areas there is relatively clear evidence for a downward trend, while in other areas rainfall appears to be stable or to be increasing. A downward trend has been established for the Sudan and Sahel zones of Africa. This has had a range of consequences, including a substantial rise in **dust-storm** activity and a severe reduction in the area and water volume of Lake Chad. By contrast, the latest analyses of summer **monsoon** rainfall for the Rajasthan Desert in India show a modest upward trend between 1901 and 1982. Data for north-east Brazil, much of Australia, and California and Arizona in the USA show no clear trend in either direction.

Attempts to reduce damage to dryland can be divided into two types: (1) technological methods; and (2) fundamental changes in societies, economies and politics. For example, a range of technological solutions is available to control blowing sand and mobile dunes (see part V, section 3). It is much more difficult, however, to make changes in the human conditions that are the real root causes of the problem. Population growth, poverty, political instability, poor planning, the attitudes of urban elites, and the prevention of traditional nomadic migration are among the long term, basic issues that need to be addressed.

Land degradation is not, however, an inevitable result of rapidly growing human populations. Excellent proof of this is provided by Tiffen et al.'s (1994) study of the semi-arid Machakos District in Kenya. Around sixty years ago this area had severe famine problems and was suffering from severe land degradation and soil erosion. Photographs from the 1930s show a gullied and impoverished landscape. Since that time the population of the district has increased more than fivefold, but the environment is now in a very much better condition than it was then. Slope **terracing** has reduced the extent and rate of soil erosion and gully formation. A fuelwood crisis has been averted by planting a large number of farmed and protected trees. In addition, agricultural output has increased. Tiffen et al. argue that high rates of population growth can be combined with sustainable environmental management. In Machakos District, the local Akamba people have proved very adaptable. Labour is plentiful, and they have invested both labour and capital in land improvement and development. They have added to their agricultural incomes by doing much more non-farm work, and the huge growth in the output of non-subsistence products has led to a development of jobs and skills in marketing and processing. In short, humans can manage the landscape to good effect even when their numbers increase.

FURTHER READING

Goudie, A. S. (ed.), 1990, *Techniques for Desert Reclamation*. Chichester: Wiley.
This edited work looks at some of the solutions that are available for dealing with the problems of desert environments.

Grainger, A., 1990, *The Threatening Desert: Controlling Desertification*. London: Earthscan.
A very readable and wide-ranging review of desertification.

Middleton, N. J., 1991, *Desertification*. Oxford: Oxford University Press.
A well-illustrated, simple introduction designed for use in secondary education.

Desertification in north central China

Deserts and 'desertified lands' cover some 1.49 million sq km of China, amounting to approximately 15.5 per cent of the total land area of the country. There are 12 named deserts within China, and it is estimated that various human and natural forces are combining to produce desertification of some 1,560 sq km per year around these deserts (see table II.2). Three main types of desertification are found in China: the spread of

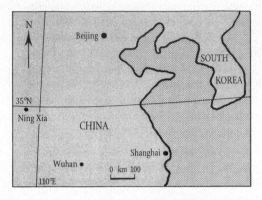

desert-like conditions on sandy **steppe**; reactivation of 'vegetated dunes' (sandy lands); and encroachment of mobile dunes on to settlements and farmland. Studies within China suggest that such desertification is a 'blistering' process, i.e. that it occurs in blister-like localized patches of rangeland away from the desert margin. These blisters then gradually grow and coalesce to produce large patches of desertified land. According to Fullen and Mitchell (1994) Chinese desertification

Plate II.2   Encroaching sand dunes on the edge of the Gobi Desert, Dunghuang, China. The dunes are invading fields and many methods are being used to try to stabilize them. (Trip/J. Batten)

Table II.2   Types, causes and extent of desertified lands in China

Causes	Area affected (sq km)	% of total desertified land
Overcultivation on steppe	44,700	25.4
Overgrazing on steppe	49,900	28.3
Overcollection of fuelwood	56,000	31.8
Misuse of water resources	14,700	8.3
Encroachment of dunes	9,400	5.5

Source: Adapted from Fullen and Mitchell (1994).

is mainly a result of land mismanagement, encouraged by climatic factors which produce droughts and encourage sand movements.

The severity of desertification and land degradation in China has prompted the Institute of Desert Research of Academia Sinica (IDRAS) to set up research into combating desertification. IDRAS has nine research stations in desertified areas at which various reclamation techniques are practised. At the Shapotou Research Station in Ningxia Autonomous Region, established in 1957 to discover methods of protecting a major railway line from sand movements, the following techniques have been used:

- planting windbreaks of pines, poplars and willows parallel to the railway line;
- levelling dunes with bulldozers;
- installing drip irrigation systems to aid topsoil development on levelled dune sands;
- constructing straw checkerboards to stabilize surfaces and encourage plants to grow on shifting dunes, to help stabilize them: this produces an artificial ecosystem on the dunes, increasing vegetation cover from less than 5 per cent to between 30 per cent and 50 per cent, and stopping dune movement.

Irrigation, land enclosure and chemical treatments are also being used in this area to help turn desertified lands into productive rangelands. According to recent studies, such reclamation efforts must be maintained and monitored over at least six years before significant improvements can be seen.

Further reading

Fullen, M. A. and Mitchell, D. J., 1994, Desertification and reclamation in north central China. *Ambio* 23, 131–5.

4 DEFORESTATION

Clearing forests is probably the most obvious way in which humans have transformed the face of the Earth. It was the prime concern of George Perkins Marsh when he wrote his pioneering book calling for the conservation of nature, *Man and Nature*, in 1864 (see part IV, section 3). Forests provide wood for construction, for shelter and for making tools. They are also a source of fuel, and, when cleared, provide land for food production. For all these reasons they have been used by humans, sometimes to the point of destruction.

Forests, however, are more than an economic resource. They play several key ecological roles. They are repositories of biodiversity (see section 10 below); they may affect regional and local climates and air quality; they play a major role in the hydrological cycle; they influence soil quality and rates of soil formation, and prevent or slow down soil erosion.

We do not have a clear view of how fast deforestation is taking place. This is partly because we have no record on a global scale of how much woodland there is today, or how much there was in the past. It is also because there are disagreements about the precise meaning of the word 'deforestation'. For example, shifting cultivators and loggers in the tropics often leave a certain proportion of forest trees standing. At what point does the proportion of trees left standing permit one to say that deforestation has taken place? Also, in some countries (e.g. India) scrub is included as forest while in others it is not.

What we do know is that deforestation has been going on for a very long time. **Pollen analysis** shows that it started in prehistoric times, in the Mesolithic (around 9,000 years ago) and **Neolithic** (around 5,000 years ago). Large tracts of Britain

had been deforested before the Romans arrived in the islands in the first century BC. Classical writers refer to the effects of fire, cutting and the destructive nibble of goats in Mediterranean lands. The Phoenicians were exporting cedars from Lebanon to the Pharoahs and to Mesopotamia as early as 4,600 years ago. A great wave of deforestation occurred in western and central Europe in medieval times. As the European empires established themselves from the sixteenth and seventeenth centuries onwards, the activities of traders and colonists caused forests to contract in North America, Australia, New Zealand and South Africa, especially in the nineteenth century. Temperate North America, which was wooded from the Atlantic coast as far west as the Mississippi River when the first Europeans arrived, lost more woodland in the following 200 years than Europe had in the previous 2,000. At the present time the humid tropics are undergoing particularly rapid deforestation. Some areas are under particularly serious threat, including South-East Asia, West Africa, Central America, Madagascar and eastern Amazonia (figure II.5).

The effects of deforestation can be seen especially vividly in the Mediterranean lands of the Old World. As Ponting (1991, p. 75) puts it:

Modern visitors regard the landscape of olive trees, vines, low bushes and strongly scented herbs as one of the main attractions of the region. It is, however, the result of massive environmental degradation brought about not by the creation of an artificial system such as irrigation but by the relentless pressure of long-term settlement and growing population. The natural vegetation of the Mediterranean area was a mixed evergreen and deciduous forest of oaks, beech, pines and cedars. This forest was cleared bit by bit for a variety of reasons – to provide land

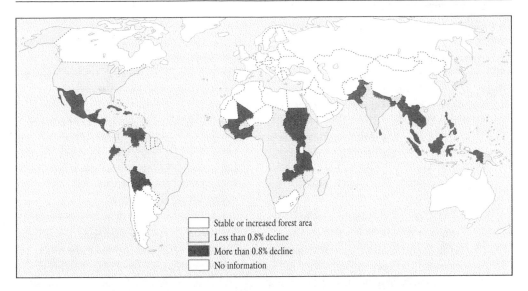

Figure II.5   Estimated annual forest change rates, 1981–1990
Source: World Resources Institute (1994), figure 7.1.

for agriculture, fuel for cooking and heating, and construction materials for houses and ships.

Other processes linked to humans, including grazing and fire, have prevented forest from returning over wide areas. In place of forest, a kind of vegetation called *maquis* has become widespread. This consists of a stand of **xerophilous** evergreen bushes and shrubs whose foliage is thick, and whose trunks are normally obscured by low-level branches. It includes such plants as holly oak (*Quercus ilex*), kermes oak (*Quercus coccifera*), tree heath (*Erica arborea*), broom heath (*Erica coparia*) and strawberry trees (*Arbutus unedo*). There is considerable evidence that *maquis* vegetation is in part adapted to, and in part a response to, fire. One effect of fire is to reduce the numbers of standard trees and to favour species which, after burning, send up suckers from ground level. Both *Quercus ilex* and *Quercus coccifera* seem to respond to fire in this way. A number of species (for example, *Cistus albidus*,

Erica arborea, Pinus halepensis) seem to be encouraged by fire. This may be because it suppresses competing plants, or perhaps because a short burst of heat encourages germination. We have already noted this happening in the chaparral of the southwest USA, an environment similar to *maquis* (see section 2 above).

Since pre-agricultural times approximately one-fifth of the world's forests have been lost. The highest losses (about a third of the total) have been in temperate areas. However, deforestation is not an unstoppable or irreversible process. For example, a 'rebirth of forest' has taken place in the USA since the 1930s and 1940s. Many forests in developed countries are slowly but steadily expanding as marginal agricultural land is abandoned. This is happening both because of replanting schemes and because of fire suppression and control (see section 2 above). Also, in some cases the extent and consequences of deforestation may have been exaggerated. A classic example of this is provided by the mountains of parts of Nepal. It was generally believed that rapid deforestation and changes in land use here had contributed to higher flood **runoffs**, floods, soil erosion and increases in river sediment loads. The effects were thought to extend as far as the Ganges Delta in Bangladesh. A detailed study by Ives and Messerli (1989), however, has cast doubt on this argument by showing that little reduction in forest cover has taken place in the Middle Mountains of Nepal since the 1930s.

Many of the phenomena noted in Nepal – flood runoffs, soil erosion, etc. – may be natural and inevitable consequences of the presence of steep mountains, rapid uplift by **tectonic** forces and monsoonal storms. Nonetheless, the loss of moist rain forests in some of the world's humid tropical regions is a very major concern. The consequences are many and serious (table II.3). The causes are also diverse and include encroaching cultivation and pastoralism (including cattle ranching), mining and hydroelectric schemes, as well as logging operations themselves.

Views vary as to the present rate of rainforest removal. Recent FAO estimates (Lanly et al., 1991) put the total annual deforestation in 1990 for 62 countries (representing some 78 per cent of the tropical forest area of the world) at 16.8 million hectares. This figure is significantly higher than the one obtained for these same countries for the period 1976–80 (9.2 million hectares per year). Myers (1992) suggests that there has been an 89 per cent increase in the tropical deforestation rate during the 1980s. This contrasts with an FAO estimate of a 59 per cent increase. Myers believes that the annual rate of loss in 1991 amounted to about 2 per cent of the total forest expanse.

Plainly, therefore, rain forests, which Myers (1990) describes as 'these most exuberant expressions of nature', are under threat. A very significant proportion of them will disappear in the next few decades unless some form of action is taken to prevent this.

Possible solutions to the tropical deforestation problem are as follows:

- *research, training and education* to give people a better understanding of how forests work and why they are important, and to change public opinion so that more people appreciate the uses and potential of forests;
- *land reform* to reduce the mounting pressures on landless peasants caused by inequalities in land ownership;
- *conservation of natural ecosystems* by setting aside areas of rain forest as National Parks or nature reserves;
- *restoration and reforestation* of damaged forests;
- *sustainable development*, namely development which, while protecting the habitat, allows a type and level of economic activity that can be sustained into the future with minimum damage to people or forest (e.g. selective logging rather than clear felling; promotion of non-tree forest products; small-scale farming in plots within the forest);
- *control of the timber trade* (e.g. by imposing heavy taxes on imported tropical forest products and outlawing the sale of tropical hardwoods from non-sustainable sources);
- *'debt-for-nature' swaps* whereby debt-ridden tropical countries set a monetary value on their ecological capital assets (in this case forests) and literally trade them for their international financial debt;
- *involvement of local peoples* in managing and developing the remaining rain forests;
- *careful control of international aid* and development funds to make sure they do not inadvertently lead to forest destruction.

Table II.3 The consequences of tropical deforestation	
Type of change	*Examples*
Reduced biological diversity	Species extinctions Reduced capacity to breed improved crop varieties Inability to make some plants economic crops Threat to production of minor forest products
Changes in local and regional environments	More soil degradation Changes in water flows from catchments Changes in buffering of water flows by wetland forests Increased sedimentation of rivers, reservoirs, etc. Possible changes in rainfall characteristics
Changes in global environments	Reduction in carbon stored in the terrestrial biota Increase in carbon dioxide content of atmosphere Changes in global temperature and rainfall patterns through greenhouse effects Other changes in global climate due to changes in land surface processes

Source: Grainger (1992).

The situation is complex but it is also urgent. No simple or single solution will be adequate. 'The time-bomb of ecological, environmental, climatic and human damage caused by deforestation continues to tick' (Park, 1992, p. 162).

FURTHER READING

Aiken, S. R. and Leigh, C. H., 1992, *Vanishing Rainforests: Their Ecological Transition in Malaysia*. Oxford: Oxford University Press.
A case study from a threatened area.

Grainger, A., 1992, *Controlling Tropical Deforestation*. London: Earthscan.
An up-to-date introduction with a global perspective.

Park, C. C., 1992, *Tropical Rainforests*. London: Routledge.
Another relatively simple introduction to many aspects of the rain-forest environment.

Williams, M., 1989, *Americans and their Forests*. Cambridge: Cambridge University Press.
A very full and scholarly discussion of the historical geography of American forests.

Managing tropical rain forest in Cameroon

Cameroon in West Africa is only the 23rd largest country on the continent, but it contains the fifth highest number of mammal and plant species, as well as populations of over 40 globally threatened animals (Alpert, 1993). It is part of an important heartland of diversity, containing many **endemic** species. In the lowland forests of Cameroon and south-east Nigeria there are over 8,000 endemic plant species, as well as endemic animals such as the Cameroon woolly bat (*Kerivoula muscilla*) and pygmy squirrel

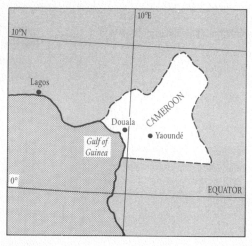

(*Myosciurus pumilio*). Lowland evergreen tropical forest covers 34 per cent of the country. Sixty per cent of this total is classed as degraded and 4 per cent as protected. According to surveys in the mid-1980s, some 17 million hectares have been deforested out of an original area of nearly 38 million hectares. In the decade 1976–86 0.6 per cent of the total forest was lost each year.

Plate II.3   The landscape of Rumsiki, Cameroon. (Panos Pictures/Victoria Keble-Williams)

Exploitation of tropical forests in this part of Africa has gradually spread inland from the west coast. Cameroon has more forests left than any coastal West African country, but less than any Central African country such as Zaire. The major cause of deforestation at the moment is felling for fuelwood and charcoal, but there are also increasing industrial demands for timber and forest products. Out of a total of over 14 million cu metres of wood produced by Cameroon in 1989–91, more than 11 million cu metres was roundwood for fuel and charcoal. Hunting is also a major threat to animal life in the tropical forests.

Cameroon established laws to manage and protect its tropical forests in 1981. This legislation decreed that 20 per cent of national territory

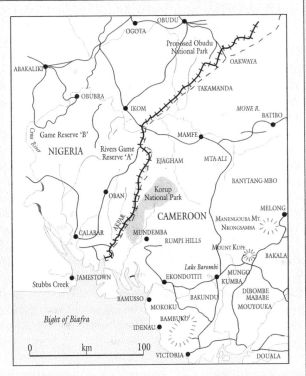

Figure II.6 National parks and reserves in Cameroon

Source: After Kingdon (1990), pp. 272–3.

should be designated as state forests. Most of these are to be productive, not protected, forests. However, several protected areas have been established within the forests: an example is the Korup National Park, which covers 1,260 sq km and has 15 staff. Maintenance is supported by the WFN (Worldwide Fund for Nature)/WWF (World Wildlife Fund) and other international bodies.

Figure II.6 illustrates the distribution of protected areas within north-west Cameroon, many of which are found in tropical forests. At present, forest reserves in Cameroon are poorly funded. This may put their long-term protection at risk.

Further reading

Alpert, P., 1993, Conserving biodiversity in Cameroon. *Ambio* 22, 44–8.

Kingdon, J., 1990, *Island Africa: The Evolution of Africa's Rare Animals and Plants*. London: Collins.

Williams, M., 1994, Forests and tree cover. In W. B. Meyer and B. L. Turner (eds), *Changes in Land Use and Land Cover: A Global Perspective*, 97–124. Cambridge: Cambridge University Press.

5   TROPICAL SECONDARY FOREST FORMATION

The clearance, cultivation and subsequent abandonment of forests in the moist tropics has resulted in the development of increasing expanses of what is called **secondary forest**. In a large and steadily increasing proportion of the tropics, secondary forests make up most or all of the remaining forest.

It is difficult to define precisely what we mean by secondary forest. Some foresters would define it as a type of forest that has suffered some form of disturbance as a result of human actions. This could be slight (e.g. hunting of animals or collection of foodstuffs) or massive (e.g. clear felling). Other foresters believe it is now useful and logical to restrict the use of the term 'secondary forest' to describe forest that has regrown after clearance.

Secondary forest development is one consequence of the practice of shifting cultivation. Peasant farmers clear small plots of just a few hectares, cultivate them for a few years, and then abandon them when soil fertility and crop yields decline. The abandoned plots are then colonized by herbs, shrubs and a canopy of pioneer trees. This kind of tree grows rapidly, needs a lot of light, and has low-density wood and sparse branching. These trees are typically short-lived, with life-spans of one or two decades. There are not many different species. As the process of succession continues, the forest progressively approaches its primary state. However, it may take 500 years or even longer for the forest to recover its full diversity of species.

Exactly how the forest recovers will depend on the degree of initial disturbance. Traditional shifting cultivation employs only small plots, so that recolonization

Plate II.4   Tropical secondary forest and slash-and-burn fields in the rainforest zone of Ghana. (Rod McIntosh)

from neighbouring primary forest is relatively easy. When larger areas are cleared, or when prolonged cultivation and frequent burning takes place (leading to severe soil degradation), the process will be much slower. However, on sites which have not been seriously damaged the biomass of leaves and fine roots (though not total biomass) is restored to that of primary forest within as little as five to ten years, by which time net primary production (NPP) is equal to or greater than that of primary forest. Thus secondary forest is probably highly effective at providing what are called 'ecosystem services' – that is, at preventing soil erosion and regulating runoff. It also has some conservation value, in that it provides a refuge for some forest fauna and a habitat for some flora.

Further Reading

Corlett, R. T., 1995, Tropical secondary forests. *Progress in Physical Geography* 19, 159–72.

Ellenberg, H., 1979, Man's influence on tropical mountain ecosystems in South America. *Journal of Ecology* 67, 401–16.

6 Grasslands and Heathlands: The Human Role

In the highlands of Africa there are large areas of what are called 'Afromontane Grasslands'. They extend as a series of 'islands' from the mountains of Ethiopia to those of the Cape area of South Africa. Are they the result mainly of forest clearance by humans in the recent past? Or are they a long-standing and probably natural component of the pattern of vegetation (Meadows and Linder, 1993)? Are they caused by frost, seasonal aridity, excessively poor soils or an intensive fire regime? This is one of the great controversies of African vegetation studies.

Almost certainly a combination of factors has given rise to these grasslands. On the one hand current land management practices, including the use of fire, prevent forest from expanding. There has undoubtedly been extensive deforestation in recent centuries. On the other hand, pollen analysis from various sites in southern Africa suggests that grassland was present in the area as long ago as 12,000 BP. This would mean that much grassland is not derived from forest through very recent human activities.

Similar arguments relate to many other of the world's great areas of grassland. Consider, for example, the savannas of tropical regions, which cover about 18 million sq km. Grasses and sedges make up most of the vegetation in savanna, although woody plants are present in varying proportions. As with most major vegetation types, a large number of interrelated factors are involved in causing savanna. It is important to distinguish clearly 'predisposing', 'causal', 'resulting' and 'maintaining' factors. For instance, around the periphery of the Amazon Basin it appears that the climate *predisposes* the vegetation toward the development of savanna rather than forest. The **geomorphological** evolution of the landscape and the formation of heavily leached old erosion surfaces may be a *causal* factor; increased **laterite** (iron crust) development a *resulting* factor; and fire, a *maintaining* factor.

Originally savanna was believed to be a predominantly natural vegetation type, developed to suit particular climatic conditions (figure II.7). It was thought that savanna is better adapted than other

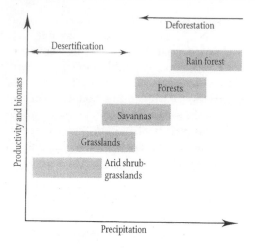

Figure II.7 An idealized relation between the biomass density and productivity of five major vegetation types (biomes) and precipitation. As precipitation increases so does productivity, and therefore biomass, with the two extremes being the low, sparsely shrubbed grasslands of the deserts and the tall, closed forests, be they tropical, temperate or boreal
Source: Graetz (1994), pp. 125–47.

plant formations to cope with the great fluctuations in rainfall during the year in the seasonal tropics. Rain forests could not resist the long winter droughts, while dry forests could not compete successfully with **perennial** grasses during the lengthy period of water surplus in the summer months.

Other workers have emphasized the importance of **edaphic** (soil) conditions. They argue that the development of savanna is encouraged by poor drainage, soils with a low water-retention capacity in the dry season, soils with a shallow **profile** due to the development of a lateritic crust, and soils with a low nutrient supply. This last condition may arise because the soil has developed on a nutrient-poor parent rock such as quartzite, or because the soil has undergone an extended period of leaching over millions of years on surfaces which

have been exposed to the elements for all that time.

A third group of researchers take the view that savannas are the product of drier conditions in former times, such as the late Pleistocene. In spite of a moistening climate in the Holocene, the savannas have been maintained by fire. They point to the fact that the patches of savanna in southern Venezuela occur within areas of forest where the levels of humidity and soil infertility are similar. This suggests that neither soil fertility, nor drainage, nor climate can be pinpointed as the cause of savanna. Moreover, the present 'islands' of savanna contain plant species which are also present elsewhere in tropical American savannas. This suggests that the 'islands' were once part of a much larger continuous expanse of savanna.

The importance of fire in creating and maintaining some savannas is suggested by the fact that many kinds of tree that grow in savanna are fire-resistant. Controlled experiments in Africa demonstrate that some tree species, such as *Burkea africana* and *Lophira lanceolata*, withstand repeated burning better than others. It has also been noted that, for example, African herdsmen and agriculturalists frequently use fire over much of tropical Africa to maintain grassland. Certainly the climate of savanna areas is conducive to fire.

Some savannas are undoubtedly natural. Pollen analysis in South America shows that savanna vegetation was present before the arrival of human civilization. Nonetheless, even natural savannas change their characteristics when subjected to human pressures. For example, many studies from warm parts of the world have shown that grass cover cannot maintain itself under heavy grazing pressure. Heavy grazing tends to remove the fuel (grass) from much of the land surface. This means that fires happen much less often, allowing trees and bushes to invade the savanna.

Plate II.5   Savanna landscape in the west Kimberley region of north-western Australia. The use of fire may be important in controlling both the nature and the distribution of this extensive biome type. (A. S. Goudie)

Whatever the factors that determine the origin of savannas, there are others that help to determine some of their particular characteristics. One example we can give of this is the role of elephants in African savannas. We do this, partly because it is a good illustration of the interdependence of vegetation and animals, and partly because if elephant numbers are reduced by human pressures, then the whole character of the savanna ecosystem may change. Elephants are what is known as a 'keystone species' because they exert a strong influence on many aspects of the environment in which they live. They diversify the ecosystems which they occupy and create a mosaic of habitats by browsing, trampling and knocking over bushes and trees. They also disperse seeds through their eating and defecating habits and maintain or create water holes by wallowing. All these

roles are of benefit to other species. Conversely, where human interference prevents elephants from moving freely within their habitats and leads to their numbers exceeding the carrying capacity of the savanna, their effect can be environmentally catastrophic. Equally, if humans reduce elephant numbers in a particular piece of savanna, the savanna may become less diverse and less open, and its water holes may silt up. This will be to the detriment of other species.

The mid-latitude grasslands (the prairies of North America, for example) are also the subject of controversy as to their origins. As we discussed in section 2 above on fire, there has been a debate as to whether the prairies are essentially the result of low precipitation and high evapotranspiration levels, or whether they result from fires.

Heathland is another fascinating vegetation type. It is characteristic of temperate oceanic conditions on acidic substrates. It is composed of ericoid (or heather-like) low shrubs, which form a closed canopy at heights usually less than 2 metres. Trees and tall shrubs are either absent altogether or scattered. Some heathlands are natural. These include areas at altitudes above the forest limit on mountains and those on exposed coasts. There are also well-documented examples of heathlands which appear naturally in the course of plant succession. This can happen, for example, where *Calluna vulgaris* (heather) replaces grasses like *Ammophila arenaria* and *Carex arenaria* on coastal dunes.

However, extensive areas of heathland also occur at low and medium altitudes on the western fringe of Europe, between Portugal and Scandinavia. The origin of these heathlands is strongly disputed. Some were once thought to have developed where there were appropriate edaphic conditions (for example, well-drained loess or very sandy, nutrient-poor soils), but pollen analysis showed that most heathlands occupy areas which were formerly tree-covered. This evidence alone did not settle the question whether the change from forest to heath was more likely to have been caused by Holocene climatic change or by human activity. However, two other factors suggest that human actions established, and then maintained, most of these heathland areas. The first of these is the presence of human artefacts and buried charcoal; the second is the fact that the replacement of forest by heath has occurred at many different times between the Neolithic and the late nineteenth century. Fire is an important management tool for heather in locations such as upland Britain, since the value of *Calluna* as a food for grazing animals increases if it is periodically burned.

The area covered by heathland in Western Europe reached a peak around 1860. Since then there has been a very rapid decline. Reductions in Britain averaged 40 per cent between 1950 and 1984, and this was a continuation of a more long-term trend. In England, the Dorset heathlands that were such a feature of Thomas Hardy's Wessex novels are now a fraction of their former size. There are many reasons for this decline. They include unsatisfactory burning practices, the removal of peat, drainage, fertilization, replacement by improved grassland, conversion to forest, and the quarrying of sand and gravel.

Thus human activities over a very long time can combine with natural changes both to produce and to remove grasslands and heathlands. Many scientific debates are continuing on how such plant communities react to stress. The box opposite gives an example from Australia.

FURTHER READING

Gimingham, C. H. and de Schmidt, I. T., 1983, Heaths and natural and semi-natural vegetation. In W. Holzner, M. J. A. Werger and I. Ikusima (eds), *Man's Impact on Vegetation*, 185–99. The Hague: Junk.
The best general review of the world's heathlands.

Harris, D. R. (ed.), 1980, *Human Ecology in Savanna Environments*. London: Academic Press.
A useful collection of papers on savannas in their human context.

Recent human impacts on subalpine grassland and heathland in Victoria, Australia

Cattle grazing began in the 1850s in the Bugong High Plains alpine grassland in what is now the Victoria Alpine National Park (created in 1980). Ever since there have been debates over the degradation of grassland and soil erosion. In 1939 there were disastrous bush fires here, and in the 1940s soil erosion became very serious as stock numbers increased. Since the 1950s there has been an overall decline of about 60 per cent in both stock numbers and the area grazed, and by 1991 only about 3,100 cattle were grazing the area between December and April.

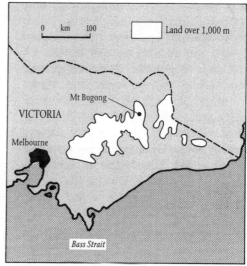

These changes in grazing densities have been echoed by an increase in the area of shrub cover. In 1945, permanent study plots were established by ecologists to monitor the changing vegetation cover on grazed and ungrazed land. The records produced from these plots over five decades enable scientists today to test the relationship between grazing, fire and the maintenance of grassland (Wahren et al., 1994). It has been suggested that cattle grazing reduces shrub cover (therefore maintaining grassland) and also fire risk. But the Bugong study does not back this up, as grazed plots have more bare patches than ungrazed (see table II.4), although by 1994 old shrubs on some ungrazed plots were beginning to die back. In this area, alpine vegetation seems slow to recover after disturbance (such as fire) and even slower where grazing is present.

Table II.4   Comparisons of percentage cover of different vegetation on grazed and ungrazed grassland plots, Bugong High Plains, Victoria, Australia, 1982–1994

Cover class[a]	1982 Ungrazed	1982 Grazed	1989 Ungrazed	1989 Grazed	1994 Ungrazed	1994 Grazed
1	76	53	71	61	66	72
2	21	31	24	20	31	17
3	3	16	5	19	3	10

[a] Cover class 1 = thick litter and dense vegetation.
  Cover class 2 = thick or thin litter, sparse vegetation.
  Cover class 3 = thin litter, sparse to no vegetation.

Source: Adapted from Wahren et al. (1994).

7   TEMPERATE FORESTS UNDER STRESS

Forest decline is an environmental issue that came to the fore in the 1980s. It has many symptoms, including the discoloration and loss of needles and leaves; reduced rates of growth; abnormal growth forms; and, in extreme cases, tree death.

Germany is probably the European country most seriously affected by forest decline. In 1985 55 per cent of the forest stands in West Germany were reported to be damaged. The decline is, however, widespread in much of Europe (see tables II.5, II.6). The process is now also undermining the health of eastern North America's high coniferous forests. In Germany it was the white fir, *Abies alba*, which was afflicted initially, but since then the symptoms have spread to at least ten other species in Europe, including Norway spruce (*Picea abies*), Scots pine (*Pinus sylvestris*), European larch (*Larix decidua*), and seven broad-leaved species.

In 1982–3 the German government adopted a comprehensive clean air legislation package. However, the data presented in table II.6 indicate that German forests are still suffering from decline. In 1994, at the Oslo international meeting, Germany agreed to reduce sulphur emissions by 83 per cent (from 1980 levels) by 2000. In 1986 the Federal Environment Ministry concluded that 'there is no single type of forest damage and no single cause. We are dealing with a highly complex phenomenon which is difficult to untangle and in which air pollutants play a decisive role.'

Many suggested explanations for this **dieback** have been put forward. They

---

Table II.5   Reported percentage of different tree species affected by forest decline in West European countries, 1984

Species	W. Germany	E. France	Switzerland	Austria	Italy (S. Tyrol)
Norway spruce	51	16	11	29	16
Silver fir	87	26	13	28	35
Scots pine	59	17	18	30	6
Beech	50	3	8	–	–
Oak	31	6	9	–	–
Others	31	6	9	–	–

Source: Goudie (1993).

Table II.6   Results from German forest damage surveys, 1986–1993: percentage of trees in classes 2–4 (i.e. defoliation > 25%) for all species

Area[a]	1986	1987	1988	1989	1990	1991	1992	1993	1994
EG			13.8	16.4	35.9				
WG	18.9	17.3	14.9	15.9	15.9				
G						25.2	26.0	24.2	24.4

[a] EG = former East Germany.
  WG = former West Germany.
  G = Germany after reunification.

Source: *Acid News*, 1995.

include poor forest management practices, ageing of stands, climatic change, severe climatic events (such as the severe summer droughts in Britain during 1976 and 1995), nutrient deficiency, viruses, fungal **pathogens** and pest infestations. However, particular attention is being paid to the role of pollution. This may take various forms, including gaseous pollutants such as sulphur dioxide (SO_2) nitrogen oxides (NOx) or ozone; acid deposition on leaves and needles; soil acidification and the associated problems of aluminium toxicity and excess leaching of nutrients (for example, magnesium); over-fertilization by deposited nitrogen; and the accumulation of trace metal or synthetic organic compounds (e.g. pesticides or **herbicides**) as a result of atmospheric deposition.

In many cases forest decline may result from a combination of stresses. For exam

ple, long-term climatic change may create a *predisposing stress* (see part I, section 6), which over a long period weakens a tree's ability to resist other forms of stress. Then there are *inciting stresses* that operate over shorter time-spans, for example, drought, severe frost or a short-lived pollution episode. These damage trees that are already weakened by the predisposing stresses. Thirdly, weakened trees are then more prone to a series of *contributing stresses* (e.g. attack by insect pests or root fungi).

There may also be different causes in different areas. Thus widespread forest death in Eastern Europe may result from high concentrations of sulphur dioxide combined with extreme winter stress. This is a much less likely explanation in Britain, where sulphur dioxide concentrations have shown a marked decrease in recent years. Indeed, Innes and Boswell (1990,

Plate II.6   Acid rain damage at Szkalrska Poreba, south-west Poland. Much of the pollution here comes from the burning of low-quality coals and lignites in Germany and the Czech Republic. (Richard Baker, Katz Pictures)

p. 46) suggest that the direct effects of gaseous pollutants in Britain appear to be very limited.

It is also important to recognize that some stresses may be especially significant for a particular tree species. In 1987 a survey of ash trees (*Fraxinus excelsior*) in Great Britain showed extensive dieback over large areas of the country. Almost one-fifth of all ash trees sampled were affected. Hull and Gibbs (1991) identified a link between dieback and the way the land is managed around the tree. They noted particularly high levels of damage in trees next to arable farmland. They suggested this might be associated with un-controlled stubble burning, the effects of drifting herbicides, and the consequences of excessive nitrate fertilizer applications to adjacent fields. However, the prime cause of dieback was seen to be the distur-bance of tree roots and the compaction of the soil by large agricultural machinery. Ash has shallow roots; if these are dam-aged repeatedly the tree's uptake of water and nutrients might be seriously reduced. Broken root surfaces would be prone to infection by pathogenic fungi.

Trees growing alongside roads which are regularly salted to reduce ice problems in cold winters may also become damaged. This may be a growing problem because the use of salt on roads has increased in recent years (figure II.8).

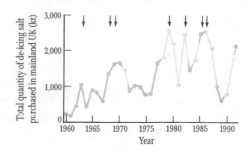

Figure II.8 Estimates of total quantity of de-icing salt purchased annually in mainland Britain during the period 1960–1991. Arrows represent years when significant crown dieback of London plane has occurred. In the early 1960s highway departments changed from using salt/abrasive mixtures to using pure rock salt. This may account for some of the increase in salt usage *Source*: Dobson (1991).

As with many environmental problems, interpretation of forest decline is hampered by a lack of long-term data and detailed surveys. Forest conditions vary from year to year in response to fluctuations in climatic stress (e.g. drought, frost, **wind throw**). This means that it is dangerous to infer long-term trends from short-term data (Innes and Boswell, 1990). The prob-lem may well have been exaggerated in the 1980s by some observers who failed to recognize that stressed trees may be a more normal phenomenon than they believed.

FURTHER READING

Boehmer-Christiansen, S. and Skea, J., 1991, *Acid Politics: Environment and Energy Policies in Britain and Germany*. London: Belhaven.

Innes, J. L., 1992, Forest decline. *Progress in Physical Geography* 16, 1–64.
An impressive overview of the competing hypotheses that have been put forward to explain forest decline.

Schulze, E.-D., Lange, O. L. and Oren, R., 1989, *Forest Decline and Air Pollution*. Ecological Studies no. 71. New York: Springer-Verlag.

Forest decline in Bavaria, Germany

Forest decline in Germany became a major environmental issue during the 1980s. Many conifers and broad-leaved trees showed signs of stress, ranging from yellowing of needles to death. In the mountains of the Fichtelgebirge in north-east Bavaria most forests at altitudes over 750 metres currently show signs of decline. By 1986, 30 per cent of Bavarian forests were classed as moderately or seriously damaged by unknown factors.

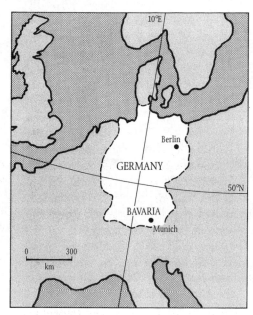

Several hypotheses have been advanced to explain the apparent decline:

- natural climatic causes and epidemics;
- direct effects of air pollution;
- mineral deficiency and imbalances as a consequence of acid deposition and soil acidification;
- a combination of some or all of the above factors.

The forests in Bavaria grow on acid, poor soils above granite and **metamorphic gneiss** and **schist** bedrocks. In the sixteenth century beech was the major species. Beech and sycamore together accounted for 60 per cent of the canopy, and fir formed the remaining 40 per cent. Over the following 400 years the forests were depleted by mining, smelting and agriculture. During the nineteenth century, reforestation took place, producing a different mix of trees. Now there are 96 per cent spruce, 2 per cent beech and 1 per cent fir (Schulze et al., 1989).

During the twentieth century, episodes of ozone, sulphur dioxide and nitrogen oxide pollution have been very severe. It now appears that air pollution, coupled with a past history of polluted air and planting, is a major problem for these very sensitive forests. Much of the pollution occurs in winter and comes from steel and chemical industries and power plants in eastern Germany and the Czech Republic. The effects of air pollution are compounded by stable winter air conditions, which encourage **temperature inversions** and the production of **smog**. Ammonia produced by nearby cattle and other farmed animals also adds to the nitrogen pollution.

8   URBAN ECOLOGY

The world is becoming increasingly urbanized. In 1980 there were 35 cities with populations of over 4 million; by 2025, 135 cities will probably have reached this size. Over the period 1950–90 the total population of the world's cities has increased tenfold, and is now more than 2 billion. Cities thus contain around half the world's population. They also contribute most global pollution. Furthermore, urban populations are concentrated into a relatively small area: for example, only 3.4 per cent of the land in the USA is urbanized. This makes the urban impact upon the environment even more intense.

The impacts of urban areas on the environment and **ecology** can be devastating. Problems have been felt for a long time in many countries where industrial cities developed early. In many less developed countries, huge expansion in population has occurred relatively recently, leading to burgeoning environmental problems.

What impacts do cities have on the environment? And how do these affect ecology? Cities do all of the following:

- produce a major demand for natural resources in the surrounding area;
- obliterate the natural hydrological system on the site of the city;
- reduce biomass and alter the species composition on the site of the city;
- produce waste products which can alter the environment in and around the city;
- create new land through **reclamation** and landfill.

Together, these impacts make up the 'ecological footprint' of a city, that is, the area affected by pollution, resource extraction, development and transport caused by the city itself. Cities demand raw materials such as timber, coal and oil: these must be extracted from the surrounding area or transported into the city. They also require agricultural products, energy and labour. As the various parts of the world become increasingly interconnected, the ecological footprints of major cities become bigger and bigger. This means that a vast proportion of the Earth's surface is being sucked into the urban system one way or another.

On the site of cities, the entire enterprise of urbanization leads to drastic changes in geomorphology, climate, hydrology and ecology. Urbanization is often seen as evidence of society's success in taming and overcoming nature. Increasing urban pollution problems, however, show that this success has been limited. Cities replace natural forests, grassland and other vegetation with vast swathes of concrete, brick and tarmac, as well as gardens, parks, ponds and derelict land. These changes in vegetation rebound upon animal life; they also affect the hydrological response. Trees, grassland and the soils in which they grow act as buffers, slowing down the movement of water through a drainage basin. As explained more fully in part IV, section 4, urbanization reduces such buffers. It accelerates and streamlines the flows of water by reducing the diffuse flow below the land surface and replacing it by flows over the surface and through pipes.

Species diversity may be increased in cities, despite the great disruption caused by building work. Gardens, parks, ponds and street plantings introduce a range of exotic plants. The urban climate also encourages growth and diversity, favouring species which tolerate warmer, less variable conditions than those found in neighbouring rural areas. The urban environment also produces behavioural changes in many animals. For example, animals which usually hibernate in winter in temperate countries can live normally throughout the year

Plate II.7   A kestrel sitting on a street lamp in a British city. These raptors, and many other organisms, have proved their ability to adapt to the urban environment. (NHPA/Michael Leach)

in large cities, where there is year-round warmth and food. Street lighting confuses birds and extends the hours of daylight for them. The vast amount of waste food found in urban areas encourages scavenging animals, such as racoons and foxes. Many urban-dwelling species have now come to be regarded as pests. Pigeons in many British and American city centres are an example: their droppings are a great nuisance. Many species arrive in cities along rivers and canals. Communications networks in general provide a major route for many animals and plants seeking to colonize new areas. In Britain, for example, mink which have escaped from fur farms are now found on urban riverbanks in Oxford.

On the other hand, the pollution and dereliction present in many cities deplete the ecology. High levels of sulphur dioxide in the atmosphere, for example, kill off lichen species growing on tree bark. Many trees themselves are threatened by air and soil pollution. Similarly, urban and industrial pollution of waterways depletes the aquatic ecology. For example, in Shanghai, China, the Huangpu River is now thought to be biologically dead as a result of the 3.4 million cu metres of industrial and domestic waste dumped in it each day. Some derelict land is highly contaminated with **heavy metals** and other toxins, thus making recolonization impossible. However, other derelict land areas provide opportunities for wildlife colonization and conservation. In Britain large areas of urban allotments (small plots of land rented out for domestic food production) are no longer cultivated, and native and exotic species are colonizing the abandoned land.

Increasingly, city dwellers are becoming

committed to improving the ecology of cities. A range of strategies is employed:

- reducing pollution to encourage desirable organisms to return;
- removing undesirable species through careful extermination programmes;
- reducing the use of lawn fertilizers and pesticides;
- planting trees in streets;
- establishing urban nature reserves, city forests and conservation areas;
- undertaking backyard composting;
- developing urban farms, thus bringing food production back into cities.

Such schemes are part of a general trend towards improving the urban environment through managing pollution and improving standards of housing and health. Sustainable development of cities is a popular phrase at present. However, vast disparities in wealth between inhabitants of different cities, and between different parts of any one city, make the goal of sustainability hard to reach. In many developing countries, squatter settlements on the outskirts of large cities are growing at an alarming rate as more and more poor inhabitants of outlying rural areas are attracted to the opportunities in cities. These settlements are very destructive of the environment, and also severely affected by environmental pollution and hazards. They usually grow up on land which is derelict because it is least suitable for development. They lack even basic services such as electricity or running water. Trees are removed so that dwellings can be put up: on steep, tropical hillslopes this can result in accelerated landslides (see part V, section 6). Wastes produced in squatter settlements cannot be removed effectively because of the lack of sanitation and services. This causes pollution of land, air and watercourses. In South Africa, for example, Soweto (which has a population of around 2.5 million according to some estimates) has horrendous air pollution from sulphur dioxide produced by coal burning, because the electricity supply is completely inadequate. The natural ecology has been wiped out, and human health is suffering.

All round the world, it is clear that the ecological impacts of cities are just one manifestation of a deep problem with present-day urbanization. As Richard Rogers, the architect, put it in 1995: 'In the beginning we built cities to overcome our environment. In the future we must build cities to nurture it.'

FURTHER READING

Bridgman, H., Warner, H., and Dodson J., 1995, *Urban Biophysical Environments*. Melbourne: Oxford University Press.
A concise introduction with an Australian flavour.

Hardoy, J. E., Mitlin, D., and Satterthwaite, D., 1992, *Environmental Problems in Third World Cities*. London: Earthscan.
Like most Earthscan books, this provides a clear introduction to the crucial issues, accompanied by many short case studies.

White, R., 1994, *Urban Environmental Management*. Chichester: Wiley.
A modern general treatment of how city environments can be managed.

Chicago's changing vegetation

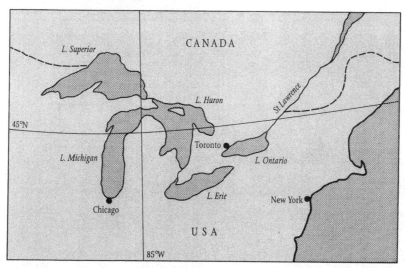

In the 1840s, before urban development really began, the flat, **glaciated** plain next to Lake Michigan in the USA was dominated by natural forest and prairie vegetation (figure II.9). Low prairie grasslands occupied most of the area. Deciduous forests of oaks (such as the Bur oak, *Quercus macrocarpa*), ashes and elms were common on sand ridges and the edges of streams. By 1860 the population of Chicago city had risen to 50,000 and by 1990 the metropolitan area contained over 8 million people.

This urban explosion has been accompanied by an almost total loss of natural vegetation, apart from some large tracts designated as forest preserves. Several direct and indirect causes of this loss of natural vegetation can be recognized. The first direct cause is the clearing of land for development. Interestingly, however, individual trees survived: some trees identified in the 1830s by the original land surveyors were still present in the 1970s (Schmid, 1975). Studies of the forest vegetation in and around Chicago show that removing the closed canopy, by creating clearings for building, favours trees which cope well in the drier and lighter conditions (such as Bur oak). Other, moisture-loving species, such as sugar maple and red oak, suffer. Indirectly, construction has disturbed the soils, affecting particularly trees such as red and white oaks. Oil spills, gas leaks, the salting of icy roads and digging to lay pipes have all had direct impacts on natural and introduced vegetation in some parts of the city.

New kinds of trees and other vegetation have been introduced into Chicago. These incomer species have had a key impact on the city's vegetation. They have also suffered from the urban environment. Interestingly, there has been an increase in the proportion of Chicago covered by trees since urbanization. This is because trees have been planted on upland sites which would naturally have been dominated by prairie.

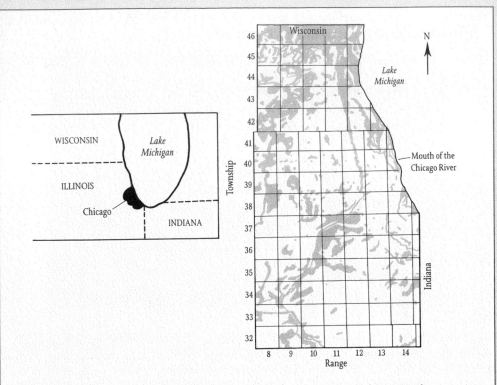

Figure II.9 The distribution of forest (black) and prairie (white) in the Chicago region during the 1840s, as recorded by the General Land Survey. Sections formed by the township/range grid are 1 mile square.
Source: Schmid (1975), figure 3.

Air pollution has had an indirect effect on vegetation. In 1913 a study found that trees in Chicago near the railway stations were affected by smoke, and the vegetation around steel mills was reduced to a few weedy annuals. More recently, the mainly deciduous trees in Chicago have shown much less damage from air pollution than the evergreens in other North American cities. A much more serious event for the urban vegetation of Chicago was the outbreak of Dutch elm disease from the late 1950s. Until 1950 American elm was the most commonly planted tree species here; Dutch elm disease destroyed the great majority of them.

Housing brings grass and shrubs, planted for decorative purposes in gardens. Chicago residents, especially in the wealthy suburbs, have planted many exotic shrub and herb species, but few native ones. Of the native plants, opportunistic herbs from flood plains have been the most successful. These plants thrive on wasteland and abandoned plots. The vegetation pattern of Chicago is now controlled by economic, social and cultural factors. The number and mix of species now vary according to the age and social characteristics of different neighbourhoods. Natural ecological factors are no longer so important as they were.

9   WETLANDS: 'THE KIDNEYS OF THE LANDSCAPE'

In the 1980s wetlands became a topic of great environmental concern. They were perceived to be vitally important ecosystems, as is made clear in the following introductory remarks to Mitsch and Gosselink's pioneering book, *Wetlands*:

> Wetlands are among the most important ecosystems on the Earth. In the great scheme of things, it was the swampy environment of the Carboniferous Period that produced and preserved many of the fossil fuels on which we now depend. On a much shorter time scale, wetlands are valuable as sources, sinks, and transformers of a multitude of chemical, biological, and genetic materials. Wetlands are sometimes described as 'the kidneys of the landscape' for the functions they perform in hydrologic and chemical cycles and as the downstream receivers of wastes from both natural and human sources. They have been found to cleanse polluted waters, prevent floods, protect shorelines, and recharge groundwater **aquifers**. Furthermore, and most important to some, wetlands play major roles in the landscape by providing unique habitats for a wide variety of flora and fauna. While the values of wetlands for fish and wildlife protection have been known for several decades, some of the other benefits have been identified more recently. (Mitsch and Gosselink, 1986, p. 3)

Wetlands are also perceived to be under threat, most notably from draining, ditching, dredging, **filling**, pollution and **channelization**. According to some sources, the world may have lost half of all its wetlands since 1900, and the USA alone has lost 54 per cent of its original wetland area,

---

Table II.7   Threats to wetlands

Source	Type	Examples
Human	Direct	Drainage for crops, timber, mosquito control
		Dredging and stream channelization
		Filling for waste disposal and land claim
		Construction of dykes, dams and sea walls for flood control and storm protection
		Discharge of materials (e.g. pesticides, nutrients from sewage, sediments) into waters and wetlands
		Mining of wetland soils for peat, coal, gravel and other minerals
Human	Indirect	Sediment diversion by dams and other structures
		Hydrological alterations by canals, roads, etc.
		Subsidence from extraction of groundwater, oil, etc.
Natural	Direct and indirect	Subsidence (including natural rise of sea level), droughts, hurricanes and other storms, erosion and biotic effects

Source: Adapted from Maltby (1986), p. 92.

Table II.8   Wetland terms and types

Name	Definition
Swamp	Wetland dominated by trees or shrubs (US definition). In Europe, a forested fen (see below) could easily be called a swamp. In some areas, wetlands dominated by reed grass are also called swamps.
Marsh	A frequently or continually inundated wetland characterized by emergent herbaceous vegetation adapted to saturated soil conditions.
Bog	A peat-accumulating wetland that has no significant inflows or outflows and supports acid-loving mosses, particularly *Sphagnum*.
Fen	A peat-accumulating wetland that receives some drainage from surrounding mineral soil and usually supports marshlike vegetation.
Peatland	A generic term for any wetland that accumulates partially decaying plant matter.
Mire	Synonymous with any peat-accumulating wetland (European definition).
Moor	Synonymous with peatland (European definition). A high moor is a raised bog, while a low moor is a peatland in a basin of depression that is not elevated above its perimeter.
Muskeg	Large expanses of peatland or bogs; particularly used in Canada and Alaska.
Bottomland	Lowlands along streams and rivers, usually on alluvial floodplains that are periodically flooded.
Wet prairie	Similar to a marsh.
Reed swamp	Marsh dominated by *Phragmites* (common reed); term used particularly in Eastern Europe.

Source: Modified from Gleick (1993), table F.1.

primarily because of agricultural developments. The pressures on wetlands are listed in table II.7.

What precisely are wetlands? There is no single, universally recognized definition, because they take a variety of forms and occur in a considerable range of conditions (table II.8). However, Maltby's (1986) definition is a useful one. He defines wetlands as 'ecosystems whose formation has been dominated by water, and whose processes and characteristics are largely controlled by water. A wetland is a place that has been wet enough for a long time to develop specially adapted vegetation and other organisms.' Wetlands therefore include areas of marsh, mire, swamp, fen, peatland or water, whether natural or artificial, per-

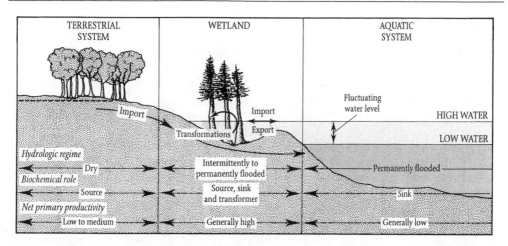

Figure II.10   Diagram showing the hydrological and ecological characteristics of wetlands, which act as ecotones between dry terrestrial ecosystems and permanently wet aquatic ecosystems
Source: Mitsch and Gosselink (1986), figure 1.4.

manent or temporary. The water may be static or flowing, fresh, brackish or salty, including marine water whose depth at low tide does not exceed 6 metres.

Wetlands cover significant areas. In all, they account for about 6 per cent of the Earth's land surface. This is not much less than the proportion of land under tropical rain forest. They also account for as much as a quarter of the Earth's total net primary production. Notable wetlands include the Everglades in Florida, the Sudd and Okavango swamps of Africa, the Fens and Broads of East Anglia in England, and the mangrove swamps of South and South-East Asia.

Wetlands are what are known as **ecotones**, that is, transitional zones. They occupy the transitional space between largely dry terrestrial systems and deep-water aquatic systems (figure II.10). This transitional position in the landscape allows them to play valuable roles, for example as **nutrient sources** or **nutrient sinks**. It also causes them to have high biodiversity, for they acquire and contain species from both terrestrial and aquatic systems.

Why are wetlands important and deserving of careful management? Because:

- they are fertile and highly productive ecosystems;
- they support fisheries of great value;
- they absorb and store carbon, which has implications for the greenhouse effect;
- they sift dissolved and suspended material from floodwaters, thereby encouraging plant growth and maintaining water quality;
- they absorb and store floodwater (thereby mitigating flood peaks) and act as barriers against storm surges, etc.;
- they are vital breeding and nursery grounds for waterfowl, animals and plants, and provide refuges in times of drought;
- they provide staple food plants (e.g. rice);
- they provide fuel (e.g. peat);
- they provide building materials (e.g. mangrove wood, reeds for thatch, etc.);
- they have recreational uses.

Because of the great value of wetlands, in 1971 many countries signed the Convention on Wetlands of International Importance especially as Waterfowl Habitat. As this was signed at Ramsar in Iran, it is often more conveniently known as the

Plate II.8   The Niger River of West Africa creates a great wetland, its so-called 'inland delta'. This photograph shows an area flooded by the annual inundation near Jenne, Mali. (Rod McIntosh)

Ramsar Convention. Those states that have signed the Convention, which now amount to over 90, agree to designate at least one of their national wetlands for inclusion in a 'List of Wetlands of International Importance'. They also agree to formulate and implement their planning so as to promote the conservation of the wetlands included in the List, to establish wetland nature reserves, and to co-operate in the management of shared wetlands and shared wetland species.

International collaboration is, of course, essential. It is no use conserving a wetland in one country to provide a refuge for species that spend one particular season of the year at that wetland, if another country destroys the refuge which they use in another season of the year. Full details of international environmental conventions of this type are listed each year in the *Green Globe Yearbook*, which is prepared by the Fridtjof Nansen Institute of Norway, and published by Oxford University Press.

FURTHER READING

Maltby, E., 1986, *Waterlogged Wealth: Why Waste the World's Wet Places?* London: Earthscan.
A very useful statement of why wetlands are important and the stresses they face.

Williams, M. (ed.), 1990, *Wetlands: A Threatened Landscape*. Oxford: Blackwell.
A more advanced collection of papers that deals with many different types of wetlands from an international perspective.

Wetlands management in the Niger Inland Delta

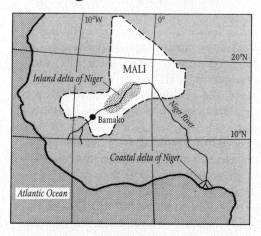

The Niger Inland Delta in Mali, Africa, is in the Sahel zone. It forms an important seasonally flooded wetland environment in an area where evaporation vastly exceeds precipitation. It covers some 20,000–30,000 sq km in the flood season and 4,000 sq km at low water and supports a population of around 550,000 people (Adams, 1993). Fishing, grazing, and cultivation of rice and sorghum are all important activities. Seventy-five per cent of the fish caught along the entire River Niger are caught here; half the total rice area in Mali is found here. Over 2 million sheep and goats and around 1 million cattle graze on the delta in the dry season. These numbers make up around 20 per cent of all these animals found in Mali.

The key to the complex and abundant agricultural production of the delta is the timing of the floods. The high flow in the delta does not coincide with the local rainfall peak. This means that there is frequent variation in environmental conditions throughout the year. The floodwaters peak between September and November and recede between December and February; rains fall between June and September; and the delta is dry between April and June. Different activities dominate the delta under these different hydrological conditions. Rice is planted as the waters rise in July and August and harvested as they recede in December to February. Sorghum is planted on the falling flood in January, and the delta is extensively used for grazing from December to July.

This wetland is also of international importance. Migrating birds visit it and the delta provides an important stop for them on the routes from the Arctic to other parts of Africa.

Failure of the rains, and alterations to the flow of the Niger River, may have serious consequences for the rich natural and human ecology of the Niger Inland Delta. Dams further up the Niger River are likely to remove about 12 per cent of inflow to the delta in a dry year, which could have impacts on fishing and agriculture. Damming of rivers tends to affect both the quality and the quantity of water, and to detract from the significant economic uses of downstream wetlands. One solution on regulated rivers may be controlled flooding, with artificial production of floodwaters from hydroelectric dams. In this way the dams are made to work *with* the natural river environment, rather than replacing it.

Further reading

Adams, W. M., 1993, Indigenous use of wetlands and sustainable development in West Africa. *Geographical Journal* 159, 209–18.

10   BIODIVERSITY AND EXTINCTIONS

What is biodiversity? It has recently been described as 'an enormous cornucopia of wild and cultivated species, diverse in form and function, with beauty and usefulness beyond the imagination' (Iltis, 1988, p. 98). Biodiversity has recently become a major environmental issue. With environments being degraded at an accelerating rate, much diversity is being irretrievably lost through the destruction of natural habitats. At the same time, science is discovering new uses for biological diversity.

The fundamental concern is the finality of the loss of biodiversity. Once a species has gone it cannot be brought back. The dodo (a bird) is dead and gone, and will never be seen again.

Biodiversity has five main aspects:

- the distribution of different kinds of ecosystems, which comprise communities of plant and animal species and the surrounding environment and which are valuable not only for the species they contain, but also in their own right;
- the total number of species in a region or area;
- the number of endemic species (species whose distribution is confined to one particular location) in an area;
- the genetic diversity of an individual species;
- the sub-populations of an individual species, that is, the different groups which represent its genetic diversity.

The Earth's genes, species and ecosystems have evolved over a period of 3,000 million years. They form the basis for human survival on the planet. However, human activities are now leading to a rapid loss of many of the components of biodiversity. Human self-interest argues that this process should be stemmed, for ecosystems play a major role in the global climate, are a source of useful products, preserve genetic strains which crop breeders use to improve cultivated varieties of plants, and conserve the soil.

We have no clear idea of the total number of species of organisms that exist on the face of the Earth. Therefore it is difficult to predict what numbers of species may be lost in the coming decades. However, according to Myers (1979, p. 31), 'during the last quarter of this century we shall witness an extinction spasm accounting for one million species.' This is a considerable proportion of the estimated number of species living in the world today, which Myers puts at between 3 and 9 million. He has calculated that from AD 1600 to 1900 humans were causing the demise of one species every four years, that from 1900 onwards the rate increased to an average of around one per year, that at present the rate is about one per day, and that within a decade we could be losing one every hour! By the end of the century, our planet could have lost anything between 20 per cent and 50 per cent of its species (Lugo, 1988). It is obvious from even this brief look at the question that the need to maintain biodiversity has become one of the crucial issues with which we must contend.

Some environments are particularly important for their species diversity. Such biodiversity 'hot spots' (figure II.11) need to be made priorities for conservation. They include coral reefs (see part VI, section 7), tropical forests (which support well over half the planet's species on only about 6 per cent of its land area), and some of the Mediterranean climate ecosystems (including the extraordinarily diverse Fynbos shrublands of the Cape region of South Africa). Some environments are crucial because of their high levels of species diversity

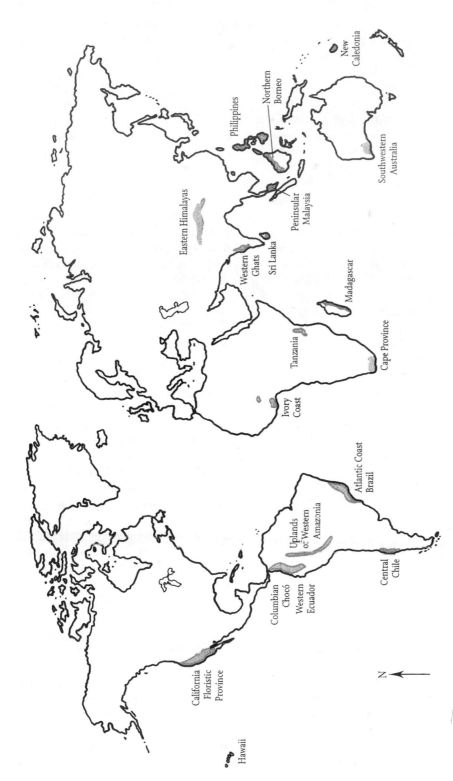

Figure II.11 Forest and heathland hot spot areas. Hot spots are habitats with many species found nowhere else and in greatest danger of extinction from human activity.
Source: Wilson (1992), pp. 262–3.

or endemic species; others are crucial because their loss would have consequences elsewhere. This applies, for example, to wetlands which provide habitats for migratory birds and produce the nutrients for many fisheries.

Reduction in habitat area can lead to a decline in the population of each species, as well as in the number of different species that the habitat can hold. Low populations make species highly vulnerable to **inbreeding**, disease, habitat alteration and environmental stress. If a species has been reduced to one population in one small area, a single fire, a single disease, the loss of a food source, or any other such 'demographic accident' can lead to extinction.

Human demographic success has produced the biodiversity crisis. As E. O. Wilson noted in his remarkable book *The Diversity of Life*:

> Human beings – mammals of the 50-kilogram weight class and members of a group, the primates, otherwise noted for scarcity – have become a hundred times more numerous than any other land animal of comparable size in the history of life. By every conceivable measure, humanity is ecologically abnormal. Our species appropriates between 20 and 40 per cent of the solar energy captured in organic material by land plants. There is no way that we can draw upon the resources of the planet to such a degree without drastically reducing the state of most other species. (Wilson, 1992, p. 272)

With the human population expected to double or treble by the middle of the twenty-first century, and the material and energy demands of developing countries likely to accelerate at an even faster rate, even less habitat will be left for other species.

What needs to be done? Wilson suggests five enterprises that need to be undertaken to save and use in perpetuity as much as possible of the Earth's diversity.

1 *Survey the world's fauna and flora* We know very little about how many species there are and even less about their qualities or where they are. Threatened habitats need to be paid particular attention.

2 *Create biological wealth* As our records of species expand, they open the way to what is called bioeconomic analysis – the broad assessment of the economic potential of entire ecosystems. An aim will be to protect ecosystems by assigning to them a future value. We need, for example, to search among wild species (possibly using ethnographic sources) for useful medical or chemical products.

3 *Promote sustainable development* As with desertification (see section 3 above), the root cause of the problem lies in society. The solution lies there as well. As Wilson (1992, p. 322) explained it: 'The rural poor of the third world are locked into a downward spiral of poverty and the destruction of diversity . . . Lacking access to markets, hammered by exploding populations, they turn increasingly to the last of the wild biological resources. They hunt out the animals within walking distance, cut forests that cannot be regrown, put their herds on any land from which they cannot be driven by force. They use domestic crops ill suited to their environment, for too many years, because they know no alternative. Their governments, lacking an adequate tax base and saddled with huge foreign debts, collaborate in the devastation of the environment.'

4 *Save what remains* Seed banks, botanical gardens, zoos and the like (the

so-called *ex situ* **methods**) may have some limited scope to preserve genetic material. However, the key issue is the preservation of natural ecosystems. We need large areas of reserves to include as many of the undisturbed habitats as possible. Priority should be given to biodiversity hot spots.

5 *Restore the wild lands* Existing ecosystems need to be salvaged and regenerated.

Plate II.9 A giant panda, *Ailuropoda melanoleuca,* feeding on bamboo at Wolong, Sichuan, China. The panda has become a symbol of the world's wildlife conservation movement. (Heather Angel)

FURTHER READING

Myers, N., 1979, *The Sinking Ark: A New Look at the Problem of Disappearing Species.* Oxford: Pergamon Press.
One of the classic statements about extinctions and biodiversity loss by one of the most persuasive environmental writers.

Wilson, E. O., 1992, *The Diversity of Life.* London: Penguin.
A beautiful piece of science writing for the lay person by a leading biologist.

Pandas, plants and parks: conserving biodiversity in China

China is both the world's most populous country and an important storehouse of global biodiversity. The country is home to around 30,000 species of plants and around 400 species of mammals. Exact statistics are hard to obtain, but we do know that there are many unique, endemic species found in China, such as the well-known giant panda (*Ailuropoda melanoleuca*). In 1965 there were 19 nature reserves covering 6,500 sq km (0.07 per cent of the total land area of China). In 1991 there were 708 reserves covering 560,000 sq km or 5.83 per cent of the total land area.

The history of nature conservation and the preservation of biodiversity in China reflects the changing social, economic and political conditions prevailing in the country. Before 1944, China had a patchy history of nature conservation as rulers established parkland, hunting grounds, gardens and temple areas. Many temple lands and sacred sites acted as biodiversity reserves.

The first modern nature reserve in China was established in 1956, in the Dinghu Subtropical Mountain Forest in Guandong Province (Freedman, 1995). The mountain was the site of an ancient Buddhist temple and so had already received much protection. Two-thirds of the reserve's 1,200 hectares had been planted with pine, or subjected to other land use modifications. In 1980 it was incorporated into the International Biosphere Network. It is now a major tourist destination, with up to 700,000 visitors a year.

Table II.9 shows that most nature reserves were not established until after 1980, when China's nature conservation laws began to multiply. By 1989, 379 vertebrate species and 389 plant species received official protection in China. By the early 1990s 13 nature reserves were devoted to the conservation and protection of the giant panda, and even more are planned.

Hunting, poaching and traditional medicine are great threats to biodiversity conservation in China. In 1990 the country was the world's largest exporter of cat and reptile skins, and live orchids. Immense indirect threats are also posed to biodiversity by the development of industry, agriculture, transport and urban areas. Although China has made great attempts to conserve biodiversity, like all countries its nature conservation programme faces many problems:

- The distribution of nature reserves is uneven (figure II.12).
- Administration is of uneven quality.
- Many nature reserves are too small to be ecologically effective.
- Nature conservation laws are not rigorously enforced.
- Environmental education in people living near nature reserves is low, and planners do not consider the economic concerns of these people sufficiently.
- Tourism has placed an additional stress on many nature reserves.

Table II.9   Nature reserves in the People's Republic of China, 1965–1991

Year	No. of reserves	Total area of reserves (000 sq km)	% of total area of country covered by reserves
1965	19	6.5	0.07
1978	34	12.6	0.13
1980	72	16	0.17
1983	262	156	1.62
1985	310	167	1.74
1987	481	237	2.47
1991	708	560	5.83

Source: Edmonds (1994), table 8.2.

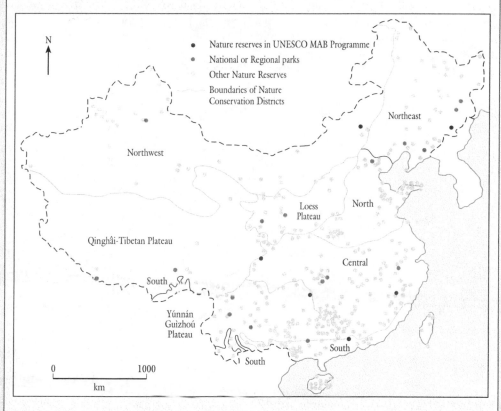

Figure II.12   The distribution of nature reserves in China
Source: Edmonds (1994), figure 8.3.

Further reading

Edmonds, R. L., 1994, *Patterns of China's Lost Harmony: A Survey of the Country's Environmental Degradation and Protection*. London: Routledge.

II INTRODUCTIONS, INVASIONS AND EXPLOSIONS

Humans are great transporters of other organisms, whether deliberately or accidentally. Thus many organisms have arrived in areas where they did not naturally occur. This applies both to plants and to animals.

Di Castri (1989) has identified three main stages in the process of biological invasions stimulated by human actions. The first stage covers several millennia up to about AD 1500. During this period, human historical events favoured invasions and migrations primarily within the Old World. The second stage began about AD 1500. At this time the exploration, discovery and colonization of new territories began in earnest, and 'the globalization of exchanges' got under way. During this phase, which lasted for about 350 years, invasions occurred from, to and within the Old World. The third stage, which only covers the last 100 to 150 years, has seen a rapidly increasing rate of exchanges and an even more extensive 'multifocal globalization', with Europe taking a less central place.

Plants that have been deliberately introduced to a new location can be divided into two groups: (1) an economic group, which consists of crops, timber trees, and cover plants for control of erosion; and (2) an ornamental or amenity group, which consists of plants introduced out of curiosity or because of their decorative value.

Plate II.10 The remarkable Fynbos heathland of the Cape Province of South Africa is rich in species, many of which are endemic. A major threat to the Fynbos is posed by the spread of invasive plants introduced from Australia. In this example it is being encroached upon from the rear by *Acacia cyclops*. (A. S. Goudie)

Table II.10   Alien plant species on oceanic islands

Island	No. of native species	No. of alien species	% of alien species in flora
New Zealand	1,200	1,700	58.6
Campbell Island	128	81	39.0
South Georgia	26	54	67.5
Kerguelen	29	33	53.2
Tristan da Cunha	70	97	58.6
Falklands	160	89	35.7
Tierra del Fuego	430	128	23.0

Source: From data in Moore (1983).

A major role in such deliberate introductions was played by botanic gardens, both those in Europe and those in the colonial territories from the sixteenth century onwards.

Plants that have been accidentally dispersed have arrived in a variety of ways: by adherence to individual people or their means of transport; among crop seed, fodder or packing materials; and as a component of transported soil, ballast, road metal or minerals.

Introduced plants are capable of invading areas to which they have been introduced, even to the extent of causing so called 'ecological explosions'. They prove to be so successful in their new habitat that they expand in range and numbers to the detriment of the native species. The same applies to introduced plant pathogens. In Britain, for instance, many elm trees have died since the 1970s because of the accidental introduction of the Dutch elm disease fungus on timber imported through certain ports in southern England. In the USA, the American chestnut was almost eliminated in less than 50 years following the introduction of the chestnut blight fungus from Asia late in the 1890s. In western Australia the great jarrah forests have been invaded and decimated by a root fungus which was probably introduced on diseased nursery material from eastern Australia.

Ocean islands have often been particularly vulnerable to invasions. The simplicity of their ecosystems inevitably leads to lower stability. Introduced species often find that the relative lack of competition enables them to spread into a wider range of habitats than they could on the continents. Moreover, because the natural species inhabiting remote islands have come to be there primarily because of their ability to disperse over large distances, they have not necessarily been dominant or even highly successful in their original continental setting. Therefore, introduced species may be more vigorous and effective. There may also be a lack of indigenous species to adapt to conditions such as bare ground caused by humans. This enables introduced weeds to establish themselves.

Table II.10 illustrates clearly how prominent alien species have become among the flora of some islands. The percentage of introduced plants varies between about one-quarter and two-thirds of the total number of species present.

Another type of ecological explosion can be caused by human-induced habitat change. Some of the most striking examples are associated with the establishment of artificial lakes behind dams in place of rivers. Riverine species which cannot cope with the changed **fluvial** conditions tend to disappear. Others that can exploit the new sources of food, and reproduce themselves under the new conditions, multiply rapidly in the absence of competition. Vegetation on land flooded as the lake waters rise decomposes to provide a rich supply of nutrients. This allows explosive outgrowth of organisms as the new lake fills. In particular, floating plants may form dense mats of vegetation, which in turn support large populations of invertebrate animals. These may cause fish to die by deoxygenating the water, and can create a serious nuisance for turbines, navigators and fishermen. On Lake Kariba in Central Africa the communities of the South American water fern (*Salvinia molesta*), bladder-wort (*Utricularia*) and the African water lettuce (*Pistia stratiotes*) grew dramatically, and on the Nile behind the Jebel Aulia Dam there was a huge increase in the number of water hyacinths (*Eichhornia crassipes*).

Various human activities, including clearing forest, cultivating, depositing rubbish and many others, have opened up a whole range of environments which are favourable to colonization by a particular group of plants. Such plants, which are not introduced intentionally, are generally thought of as 'weeds'.

Animals have been deliberately introduced to new areas for many reasons: for food, for sport, for revenue, for sentiment, for control of other pests, and for aesthetic purposes. Such deliberate actions probably account, for instance, for the widespread distribution of trout.

There have also been many accidental introductions, especially since the development of ocean-going vessels. These are becoming more frequent, for whereas in the eighteenth century there were few ocean-going vessels of more than 300 tonnes, today there are thousands. Because of this, in the words of C. S. Elton (1958, p. 31), 'we are seeing one of the great historical convulsions in the world's fauna and flora.' Indeed, many animals are introduced with vegetable products, for 'just as trade followed the flag, so animals have followed the plants'.

A recent example of the spread of an introduced insect in the Americas is provided by the Africanized honey bee. A number of these were brought to Brazil from South Africa in 1957 as an experiment, and some escaped. Since then they have moved northwards to Central America and Texas (figure II.13), spreading at a rate of 300–500 km per year, and competing with established populations of European honey bees.

Some animals arrive accidentally with other beasts that are imported deliberately. In northern Australia, for instance, water buffalo were introduced. They brought with them their own bloodsucking fly, a species which bred in cattle dung and transmitted an organism sometimes fatal to cattle. Australia's native dung beetles, accustomed only to the small sheep-like pellets of the grazing **marsupials**, could not tackle the large dung pats of the buffalo. Thus, untouched pats abounded and the flies were able to breed undisturbed. Eventually African dung beetles were introduced to compete with the flies.

Domesticated plants have, in most cases, been unable to survive without human help. The same is not so true of domesticated animals. There are a great many examples of cattle, horses, donkeys and goats which have effectively adapted to new environments and have become virtually wild (**feral**). Frequently, pigs and rabbits that have established themselves in this way have ousted native animals. Feral animals may also, particularly on ocean islands, cause desertification. Feral goats, for example, have degraded the Channel Islands off the California coast.

Figure II.13 The spread of the Africanized honey bee in the Americas between 1957 (when it was introduced into Brazil) and 1990
Source: Modified after Texas Agricultural Experiment Station, in *Christian Science Monitor*, September 1991.

Aquatic life can be spread accidentally through human alteration of waterways and by the construction of canals, which enables organisms to spread from one sea or one lake to another. This process is called **Lessepsian migration** after the name of the man who built the Suez Canal. The construction of that great waterway has enabled the exchange of animals between the Red Sea and the eastern Mediterranean. The migrants include a menacing jellyfish which has now invaded beaches on the eastern shore of the Mediterranean.

FURTHER READING

Drake, J. A. (ed.), 1989, *Biological Invasions: A Global Perspective*. Chichester: Wiley. An advanced collection of edited papers.

Elton, C. S., 1958, *The Ecology of Invasions by Plants and Animals*. London: Methuen. The classic monograph on this theme.

Alien plant species invading Kakadu National Park, Australia

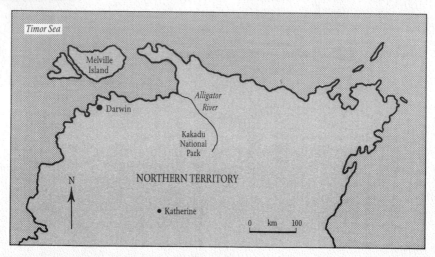

The Kakadu National Park is a UNESCO World Heritage Site in the monsoonal tropical north of Australia, containing most of the catchment of the South Alligator River. The natural vegetation is mainly savanna woodland and open forest dominated by eucalyptus. There are also extensive **alluvial floodplains**, seasonally water-covered, where herbaceous wetland vegetation grows. Out of 1,526 plant species found here, some 5.8 per cent (89 species) are considered to be invasive (Corrie and Werner, 1993). Most of these are weedy annuals from the New World tropics. Although this percentage of introduced species is low compared with the figure for the whole of Australia (10 per cent of all plant species are invasive), it is clearly a cause of worry for a nature reserve which is attracting an increasing number of visitors.

Invasions have increased by an average of 1.6 species per year since 1948 as tourism and mining have increased, bringing in more habitat disturbance. Most alien species are found around camp-sites, car parks, roads and mines. One of the biggest problems is a fast-growing shrub (*Mimosa pigra*). This plant was introduced deliberately into the Northern Territory from South America, and was not realized to be a serious nuisance until around 80 years later. It has spread over the alluvial floodplains, changing herbaceous swamps into shrublands. This, in turn, affects wildlife. There has been a major effort to control the plant.

There are also many other indirect ways in which alien plant species are spread here. Feral water buffaloes, for example, make a major contribution to invasions near floodplains as they disturb the ground.

Further reading

Kirkpatrick, J., 1994, *A Continent Transformed*. Melbourne: Oxford University Press. A concise discussion of human impacts on the natural vegetation of Australia.

12   HABITAT LOSS AND FRAGMENTATION

One of the consequences of human activities is that many natural habitats become reduced in extent and also become fragmented into isolated patches. Figure II.14 shows how both these processes have occurred in the forest cover of a part of central England in the last 1,500 years. Whereas at the end of Romano-British times (AD 400) there were still large expanses of forest, there are now only very small 'islands' of forest in a sea of agricultural land.

Certain types of habitat may be lost because of changes in agricultural practices. In Britain, for example, the botanical diversity of much pastureland has been

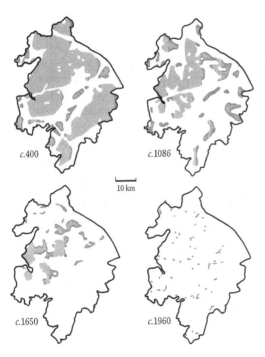

Figure II.14   Forest fragmentation in Warwickshire, England, from AD 400 to 1960. Forested areas are shown in black.
Source: Wilcove et al. (1986), figure 1.

reduced as many old meadows have been replaced with fields planted solely with grass (leys), or treated with selective herbicides and fertilizers. This treatment can take out of the habitat some of the basic requirements essential for many species. For example, the larva of the common blue butterfly (*Polyommatus icarus*) feeds upon bird's-foot trefoil (*Lotus corniculatus*). This plant disappears when pasture is ploughed and converted into a grass field, or when it is treated with a selective herbicide. Once the plant has gone, the butterfly vanishes too because it is not adapted to feeding on the plants grown in leys of improved pasture. Likewise, numbers of the large blue butterfly (*Maculinea arion*) have decreased in Britain. Its larvae live solely on the wild thyme (*Thymus drucei*), a plant which thrives on close-cropped grassland. Since the decimation of the rabbit by myxomatosis, conditions for the thyme have been less favourable, so that both the thyme and the large blue butterfly have declined.

Another major land use change of recent decades has been the replacement of natural oak-dominated woodlands in Britain and elsewhere by conifer plantations. This also has implications for wildlife. It has been estimated that where this change has taken place the number of species of birds found has been approximately halved. Likewise, the replacement of upland sheep walks with conifer plantations in southern Scotland and northern England has led to a sharp decline in numbers of ravens. The raven (*Corvus corax*) feeds on carrion, much of which it obtains from open sheep country. Other birds that have suffered from moorland areas being planted with forest trees are several types of wader, the golden eagle, peregrine falcons and buzzards.

Many species of birds in Britain have declined in numbers over the last two decades because of habitat changes resulting from more intensive farming methods. These include no longer leaving fallows, less mixed farming, new crops, modern farm management, use of biocides, and hedgerow removal.

Reducing the areas of land covered by particular habitats has a direct impact on the fortunes of species. It is useful to see the remaining fragments of habitat as 'islands'. We know from many of the classic studies in true island biogeography that the number of species living at a particular location is related to its area. Islands support fewer species than do similar areas of mainland, and small islands have fewer species than do large ones. Thus it may well follow that if humans destroy the greater part of a vast belt of natural forest, leaving just a small reserve, initially it will be 'supersaturated' with species, containing more than is appropriate to its area under natural conditons. Since there will be fewer individuals of each of the species living in the forest now, the extinction rate will increase and the number of species will decline. For this reason it is a sound principle to make reserves as large as possible. A large reserve will support more species by allowing the existence of larger populations and keeping extinction rates lower. Size, of course, is not everything and other factors such as the shape of reserves and the existence of links between reserves are also important.

Reduction in area of habitat leads to reduction in numbers of organisms. This in turn can lead to genetic impoverishment through inbreeding, with particularly marked effect on reproductive performance. Inbreeding degeneration is, however, not the only effect of small population size. In the longer term, the depletion of genetic variety is more serious, since it reduces the capacity for adaptive change. It is therefore very important to provide enough space, especially for those animals that require large expanses of territory. For

example, the population density of the wolf is about one adult per 20 sq km, and it has been calculated that for a viable population to exist, one might need 600 individuals ranging over an area of 12,000 sq km. The significance of this is apparent when one realizes that most nature reserves are small: 93 per cent of the world's national parks and reserves have an area less than 5,000 sq km, and 78 per cent cover less than 1,000 sq km.

Habitat fragmentation has some other major effects. One of these is loss of habitat heterogeneity. In other words, individual fragments may lack the full range of different habitats found in the original block. For instance, a small patch of wood may not contain a reliable water supply. Likewise, some species – certain **amphibians**, for example – require two or more habitat types. Habitat fragmentation may make it impossible for these animals to move between habitats.

A second effect of fragmentation is that the new landscape that replaces the original habitat, such as human settlements or agricultural land, may act as a barrier, preventing colonization and interchange between groups. Also, the new landscapes may enable populations to build up of animals that are harmful to species within these fragments.

A third consequence of fragmentation is what are called 'edge effects'. Some animals do well in edge habitats, that is, the boundary areas around the rim of the 'island', but others suffer. For example, many nest predators occur in higher densities around forest edges.

A fourth effect is 'secondary extinctions'. Fragmentation disrupts many of the important ecological interactions of a community. For example, small woodland 'islands' in the eastern USA contain few if any of the large predators (e.g. mountain lions) that would normally regulate the number of smaller omnivorous species (e.g.

racoon). These **omnivores** can thus prey unhindered upon the eggs and young of the forest songbirds, and may wipe them out.

FURTHER READING

Wilcove, D. S., McLellan, C. H. and Dobson, A. P., 1986, Habitat fragmentation in the temperate zone. In M. E. Soulé (ed.), *Conservation Biology: The Science of Scarcity and Diversity*, pp. 251–6. Sunderland, Mass.: Sinauer Associates.
A short but useful chapter in an advanced book.

Plate II.11   A flock of Lesser Snow Geese. (NHPA/Robert Erwin)

Texas Gulf coast habitat changes and the Lesser Snow Goose

The changing fortunes of the Lesser Snow Goose (*Chen caerulescens caerulescens*) population in Texas show interesting links between habitat changes and wildlife. Presently, around 600,000–850,000 Lesser Snow Geese winter here every year (Robertson and Slack, 1995). Until the 1920s the Lesser Snow Geese wintered mainly on coastal marshes, but now they are found on the inland prairie as well.

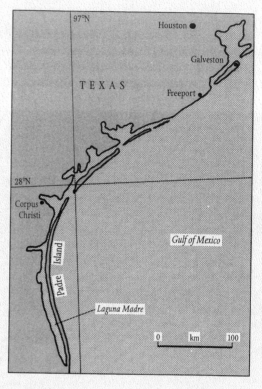

During the twentieth century, the Texas Gulf coast has seen the rise of petroleum refining and oil extraction industries, coupled with the spread of rice cultivation and a boom in population. Nearly 50 per cent of the entire USA's chemical production is based in the Houston area, and 73 per cent of the US petroleum industry is there. Rice cultivation peaked at 254,800 hectares in 1954, and now covers an area of around 141,000 hectares. These diverse changes to the landscape have caused some areas to become more suitable for the Lesser Snow Goose, while other areas have become less attractive in terms of availability of food and water.

The wintering grounds of the Lesser Snow Goose spread to the prairies between the 1920s and the 1950s following the spread of rice cultivation, although the movement of the birds lagged behind the expansion of the ricefields by some years. These changes may also have been encouraged by alterations to the coastal marsh areas, as urban and industrial development from the 1940s onwards led to marsh drainage and pollution.

The changes in wintering range were accompanied by a growth in population numbers: the Lesser Snow Goose population peaked at around 813,000 in the early 1980s. Since then, numbers have declined in association with declining rice production (the area sown with rice declined by a third from 1978 to 1991). This decline in rice cultivation was in turn related to the lack of federal price supports for rice growers, which made other crops more economically viable.

An airport planned for construction on Katy Plains would affect 1,168 hectares directly and 16,200 hectares indirectly. This project will have further impacts on the distribution and population numbers of the Texan Lesser Snow Geese.

13   EXTINCTIONS IN THE PAST

Extinctions are nothing new. They are a part of evolution, and spasms of extinction have recurred through geological time. There have been five major mass global extinctions over the last 600 million years (figure II.15). The last of these occurred at the boundary between the Cretaceous and Tertiary periods about 66 million years ago. This was when the extinction of the dinosaurs took place, possibly because of the environmental impact of a massive meteorite crashing into the Earth, or perhaps because of some major volcanic eruptions. The other mass extinctions took place in earlier periods: the Ordovician (440 million years ago), the Devonian (365 million years ago), the Permian (245 million years ago), and the Triassic (210 million years ago).

We are now living in a sixth spasm of mass global extinction. This started towards the end of the Ice Age (round about 11,000 years ago) and is accelerating at the present time. Humans are implicated in this sixth spasm, though for prehistoric times there is a major controversy as to whether the wave of extinctions might have a natural (i.e. essentially climatic) cause.

We have discussed present-day extinctions and their causes in section 10 above on biodiversity. In this section we will explore the role of our prehistoric forebears in causing the decline and extinction of many species of animal.

Over the last 30 years Paul Martin and co-workers have argued that Late Pleistocene extinctions closely followed the chronology of the spread of prehistoric human cultures and the development of big-game hunting technology. They would argue that there are no known continents or islands in which accelerated extinction definitely pre-dates the arrival of substantial numbers of humans. They would also argue that the temporal pattern of extinctions of large land mammals (the **megafauna**) follows in the footsteps of Stone Age humans. They suggest that

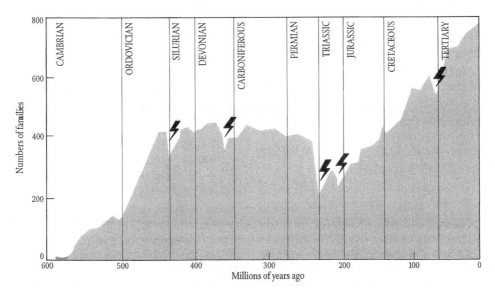

Figure II.15   Graph showing the five mass global extinctions of marine organisms (indicated by lightning flashes)
Source: Wilson (1992).

Plate II.12   A reconstruction of mammoth being hunted in Europe at the end of the Ice Age. Mammoths were one of the megafauna that became extinct at the transition from the Pleistocene to the Holocene. Was climatic change the cause or the hunting activities of our ancestors? (Natural History Museum, London)

Africa and parts of southern Asia were first affected in this way, with substantial losses around 200,000 years ago. North and South America were stripped of large herbivores between 12,000 and 10,000 years ago. Extinctions extended into the Holocene (i.e. the last 10,000 years or so) on ocean islands where humans arrived late on the scene (figure II.16).

There were three main types of human pressure involved in what is sometimes called 'Pleistocene overkill':

- the 'blitzkrieg effect', when human populations with big-game hunting technology spread rapidly so that animal populations decline very quickly;
- the 'innovation effect', when long-established human population groups adopt new hunting technologies and wipe out fauna that have already been stressed by climatic changes;

- the 'attrition effect', when extinction takes place relatively slowly after a long history of human activity because of loss of habitat and competition for resources.

What are the arguments that can be marshalled in favour of this **anthropogenic** hypothesis? First, in areas like the High Plains of America, the first massive extinctions appear to coincide with the arrival of humans who were numerous enough and who had sufficient technological skills to be able to kill large numbers of animals. Secondly, the vast number of bones at some Late Pleistocene archaeological sites attests to the efficiency of the more advanced Stone Age hunters. Thirdly, many animals unfamiliar with people are remarkably tame and naïve in their presence, rendering them easy prey. Fourthly, in addition to hunting animals to death,

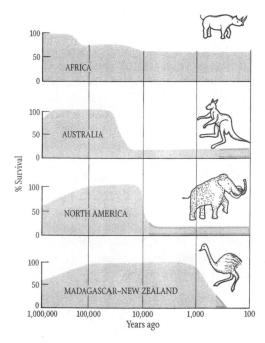

Figure II.16 The percentage survival of large animals and flightless birds over the last million years in four different areas. The extinction of these organisms coincided closely with the arrival of humans in North America, Madagascar and New Zealand, and less decisively in Australia. In Africa, where humans and animals evolved together for millions of years, the damage was less severe.
Source: Wilson (1992).

humans may also have competed with them for particular food or water supplies. Fifthly, the supposed extinction of the larger rather than the smaller mammals could be related to the effects of human predation. Large mammals have small numbers of offspring, long gestation periods, and long periods before maturity is reached. This means that populations of these animals can survive only a very low rate of slaughter, even against primitive hunters.

In addition, certain objections have been levelled against the climatic change model

which tend to support the anthropogenic model. It has been suggested, for instance, that changes in climatic zones are generally gradual enough to allow beasts to follow the shifting vegetation and climatic zones of their choice. Similar environments are available in North America today as were present, in different locations and in different proportions, during Late Pleistocene times. Secondly, it can be argued that the climatic changes associated with the multiple glaciations, **interglacials**, **pluvials** and **interpluvials** earlier in the Ice Age do not seem to have caused the same striking degree of species elimination as the changes in the Late Pleistocene. A third difficulty with the climatic cause theory is that animals like the mammoth occupied a broad range of habitats from Arctic to tropical latitudes, so that it is unlikely that all would perish as a result of a climatic change.

However, there is some support for the alternative climatic hypothesis, namely that rapid and substantial climatic change at the end of the last Ice Age led to the extinction of the great mammals like the mammoth. The migration of animals in response to the rapid climatic change at the end of the Pleistocene could be halted by geographical barriers such as high mountain ranges or seas. According to this point of view, Africa is relatively rich in big mammalian fauna because the African biota is not, or was not, greatly restricted by any insuperable geographical barrier.

Another way in which climatic change could cause extinction is through its influence on the spread of disease. It has been suggested that during glacials animals would be split into separate groups cut off from one another by ice sheets. These isolated groups might lose immunity to certain diseases to which they were no longer exposed. Then, as the ice melted (before 11,000 years ago in many areas), contacts between groups would once again be made, enabling any diseases to which immunity had been lost to spread rapidly.

It has also been noted recently that in some areas it was not only the great megafauna that became extinct. Some small animals and birds that would not have been hunted by humans also died out. Moreover, as the **radiocarbon dates** for early societies in some countries like Australia and Brazil are pushed back, it becomes increasingly clear that humans and several species of megafauna were living together for quite long periods. This is undermining the idea of rapid overkill. Also, if humans were primarily responsible for the waves of extinction, how does one explain the survival of many big game species in North America well into the nineteenth century?

The Late Pleistocene extinctions may, of course, have been caused by both climatic and anthropogenic mechanisms, or by a combination of the two types. For example, animal populations reduced and stressed by climatic change would be more vulnerable to increasing levels of human predation. Nonetheless, the rapidity with which extinctions took place in Madagascar, New Zealand and the Pacific islands after they were first settled in the Holocene is striking evidence of how even quite small numbers of technologically not very advanced people can cause major environmental change.

Further Reading

Ehrlich, P. R. and Ehrlich, A. H., 1982, *Extinction*. London: Gollancz.
An accessible treatment for the general reader.

Martin, P. S. and Klein, R. G., 1984, *Pleistocene Extinctions*. Tucson: University of Arizona Press.
A massive advanced tome from two of the leading scientists involved in the study of the possible role of humans in causing extinctions in prehistory.

14   Biotechnology, Genetic Engineering and the Environment

Biotechnology is the manipulation of living organisms and their components (e.g. genes or gene components) for specific tasks. Genetic engineering is one form of biotechnology, involving the isolation of genes and gene components that confer desired traits and their transfer between species. It is also sometimes called recombinant **DNA** technology. This branch of science has now reached a level where it is possible to transfer genes between unrelated species or types of organisms.

There are many applications of biotechnology that are of environmental relevance: in agriculture, resource recovery and recycling, pollution abatement, and the production of renewable energy resources.

In agriculture, biotechnology can help to maximize energy and nutrient flows, for example, by increasing crop yield and by engineering resistance to disease, insects and herbicides. Nitrate levels can be enhanced by seeding the ground with nitrogen-fixing bacteria. Stress tolerance (e.g. to frost) can be engineered.

Biotechnology is also being developed to undertake the recovery of resources. Mineral ores can be recovered through a process called 'biomining', which exploits the ability of specific types of bacteria to obtain their energy supply by breaking down certain types of ore-bearing deposits. Certain micro-organisms can be employed to scavenge metals from wastewater so that the metals can be re-used.

Biotechnology can also contribute to pollution abatement. 'Biosensors' are organisms that can be used to identify critical levels of pollution. Other organisms can be used to extract pollutants such as heavy metals from wastewater, to neutralize hazardous substances in the environment (**bioremediation**), or to break down sewage.

Another use of biotechnology is to produce renewable energy resources. For example, it is possible to extract alcohol from some plants; this can be used as fuel for automobiles. Protein-rich animal feeds can be obtained by using algae, fungi (including yeasts) and some bacteria to produce cellular protein from energy and nutrient sources such as carbon dioxide, methanol, ethanol, sugars and carbohydrates.

Biotechnology is potentially of enormous value and it may have many environmental benefits. For example, the environmental advantages of using biotechnology in agriculture include:

- reduced need for fuel;
- reduced use of pesticides;
- reduced use of artificial fertilizer, thus also lessening pollution by phosphates and nitrates;
- increased food supply, which could lead to less pressure on marginal lands and on remaining natural ecosystems.

Similar types of advantages can apply to the other uses of biotechnology which we have described.

On the other hand, there are possible disadvantages. These include:

- the potential to create invasive organisms – as, for instance, when genes escape into the wild relative of an engineered crop, creating potential pests;
- the potential to create organisms which are toxic or contain toxic components;
- the potential to create organisms, especially bacteria, that could profoundly alter the nature of global biogeochemical cycles.

FURTHER READING

Mannion, A. M., 1991, *Global Environmental Change*. Harlow: Longman.
A very general but useful treatment of all aspects of global change, both natural and anthropogenic.

Mannion, A. M., 1995, *Agriculture and Environmental Change*. Chichester: Wiley.
A more detailed treatment, by the same author, of biotechnology as one aspect of the agricultural impact on the environment.

15   CONCLUSIONS

In this chapter we have demonstrated that humans have had effects on the biosphere for a very long time. For many good reasons, our early ancestors developed the use of fire. This powerful technological tool has had many positive ecological consequences. It may also have had a major effect on some of the world's biomes and vegetation types, including savannas and Mediterranean shrublands. The manage-

ment of fire is an important tool for the management of some major environments. As the Yellowstone study has shown, fire suppression policies can have adverse effects.

Other major changes in the state of the world's biomes include desertification and deforestation. Both phenomena are difficult to define and to quantify. There are various ways in which desert margins and rain forests can be managed so that these processes can be kept under control. Even secondary forests, which result from human use of tropical moist forests, have positive value.

With many such changes, however, we have to recognize that very many processes, both anthropogenic and natural, may have played an important role. This is evident from a consideration of the origin of tropical savannas, heathlands and mid-latitude grasslands. Indeed, we have seen recently how complex causes can be in the case of forest decline in Europe. As we point out, there is no single type of forest damage and no single cause. Equally, we should not necessarily equate urbanization with a reduction in biodiversity. The growth of cities, as illustrated by Chicago, has major ecological consequences, but not all of them have negative impacts.

Nonetheless, there are some major habitats and particular habitat types that deserve particular attention and protection because of their importance for the preservation of biodiversity. These include wetlands and other crucial 'ecological hot spots', such as the Fynbos heathlands of southern Africa, or the forests inhabited by the Giant Panda in China. Many habitats are being considerably modified by the spread of organisms introduced by humans. These organisms may then invade susceptible habitats, of which oceanic islands are a notable example. Many other habitats are being greatly reduced in area and continuity. This creation of small 'islands' of habitat increases the likelihood of species extinctions. Extinction is an irreversible process which results from both natural and anthropogenic causes. It is one of the great challenges we face in coming decades.

We are entering a new era in the human manipulation of the biosphere. Biotechnology and genetic engineering both offer great opportunities and raise a great need for prudence.

The many case studies discussed in this part of the book show how complex human impacts on the biosphere are; how science cannot, as yet, answer all the questions; and how the many different pressures on human societies affect the ways in which they use and abuse the resources of the biosphere.

KEY TERMS AND CONCEPTS

biodiversity
biodiversity hot spots
biomass burning
biotechnology
deforestation
desertification
ecological explosion
ecosystem services
ecotones
edaphic condition
fire
fire suppression
forest decline and dieback
habitat
heathland

invasions
keystone species
Lessepsian migration
megafauna
overcultivation
overgrazing
Pleistocene overkill
prescribed burning
salinization
savanna
secondary forest
species diversity
urban ecology
wetlands
wilderness

POINTS FOR REVIEW

What do you understand by the term 'biosphere'?

Why was fire one of humankind's first technological achievements?

Should fires be suppressed?

How would you identify if desertification was taking place?

How might you aim to reduce the effects of desertification?

In what ways might tropical deforestation rates be reduced?

Assess the role of predisposing, causal, resulting and maintaining factors in the development of grasslands, savannas and heathlands.

Discuss the many different factors that could account for forest decline.

What characteristics of cities determine their impact on the environment?

Why and how should wetlands be conserved?

What do you understand by the term biodiversity?

What arguments would you use to support the view that biodiversity should be maintained?

Why should we be interested in ecological invasions and explosions?

What do you understand by the term 'habitat'?

Did climatic change or human impact cause Pleistocene extinctions?

Consider the potential role of biotechnology in environmental protection and degradation.

PART III

The Atmosphere

1   INTRODUCTION

When the great American geographer and conservationist, George Perkins Marsh, wrote *Man and Nature* in 1864 (see part IV, section 3), in which he surveyed the ways in which humankind had transformed the Earth's surface, he scarcely considered the various ways in which humans might affect the quality of the atmosphere and the nature of the Earth's climate. However, just over 100 years later, it is this very area that is the cause of greatest concern to many scientists and to others interested in environmental matters. To be sure, local air pollution was a major concern at the time Marsh wrote; but, for the most part, broader-scale human impacts on the atmosphere and climate were not given very much attention.

Since the mid-nineteenth century, when Marsh wrote his book, world industrial production and energy consumption have accelerated dramatically. All sorts of new technologies, including notably the internal combustion engine, have been introduced. As a consequence, a cocktail of gases that is growing in quantity and variety has been emitted into the atmosphere. This has created problems of poor air quality, which can affect not only human health, but also the state of whole ecosystems (for example by 'acid rain') and also of global climatic conditions (for example by the greenhouse effect). It is also apparent that changes in land use (such as deforestation), of the type discussed in part II, are causing changes at the Earth's surface which may have impacts on the climate. Great uncertainty still surrounds many of these issues, but there is no denying that matters such as **global warming**, ozone depletion and acid rain have very great implications that deserve intensive study.

Recent years have seen a great concentration of interest in the so-called 'greenhouse effect' (section 4 below) and the role that gases like carbon dioxide, methane and nitrous oxide (the 'greenhouse gases') play in global warming. There are, however, other mechanisms by which humans may cause global or regional climatic change. They are not yet fully understood, and in the long term they may not have so great an impact as the greenhouse gases. Nevertheless, they may have a significant role to play. In some cases, moreover, they could serve to counteract the greenhouse effect. In certain specific localities they may already be producing a decipherable climatic trend.

The mechanisms so far identified that may be related to human influences on global and regional climates, and their main effects, can be summarized as follows:

- *Gas emissions*
  Carbon dioxide
  Methane ⎫ Greenhouse
  Chlorofluorocarbons ⎬ gases
  Nitrous oxide ⎭
- ***Aerosol generation***
  Dust
  Smoke
  Sulphates
- *Thermal pollution*
  Urban heat generation
- *Albedo change*
  Dust addition to ice caps
  Deforestation and afforestation
  Overgrazing
  Extension of irrigation
- *Alteration of water flow in rivers and oceans*
- *Water vapour change*
  Deforestation
  Irrigation

2   ANTHROPOGENIC CLIMATE CHANGE: THE ROLE OF AEROSOLS

Let us first consider the possible effects of aerosols. An aerosol is defined as 'an intimate mixture of two substances, one of which is in the liquid or solid state dispersed uniformly within a gas. The term is normally used to describe smoke, condensation nuclei, freezing nuclei or fog contained within the atmosphere, or other pollutants such as droplets containing sulphur dioxide or nitrogen dioxide' (*Encyclopaedic Dictionary of Physical Geography*, 1985, p. 6). Many atmospheric aerosols (e.g. those derived from volcanoes, sea spray or natural fires) were not placed there by humans. However, humans have become increasingly capable of adding various aerosols into the air. For example, one consequence of the industrial revolution has been the emission of hugely increased quantities of dust or smoke particles into the lower atmosphere from industrial sources. These could influence global or regional temperatures through their impact on the scattering and absorption of solar **radiation**.

The exact effects of aerosols in the atmosphere are still not clear, however. Whether added aerosols cause heating or cooling of the Earth and atmosphere systems depends not only on their intrinsic absorption and **backscatter** characteristics, but also on their location in the atmosphere with respect to such variables as cloud cover, cloud reflectivity and underlying surface reflectivity. So, for example, over ice caps 'grey' aerosol particles would warm the atmosphere because they would be less reflective than the white snow surfaces beneath. Over a darker surface, on the other hand, they would reflect a greater amount of radiation, leading to cooling. Thus it is difficult to assess precisely the effects of increased aerosol content in the atmosphere.

Uncertainty is heightened because of the two contrasting tendencies of dust: the backscattering effect, producing cooling, and the thermal-blanketing effect, causing warming. In the second of these, dust absorbs some of the Earth's thermal radiation that would otherwise escape to space, and then re-radiates a portion of this back to the land surface, raising surface temperatures. Natural dust from volcanic emissions tends to enter the **stratosphere** (where backscattering and cooling are the main con-

sequences), while anthropogenic dust more frequently occurs in the lower levels of the atmosphere, where it could cause thermal blanketing and warming.

Industrialization is not the only source of particles in the atmosphere, nor is a change in temperature the only possible consequence. Intensive agricultural exploitation of desert margins, such as in Rajasthan, India, can create a dust pall in the atmosphere by exposing larger areas of surface materials to deflation in dust storms. This dust pall can change atmospheric temperature enough to cause a reduction in **convection**, and thus in rainfall. Observations of dust levels over the Atlantic during the drought years of the late 1960s and early 1970s in the Sahel suggest that the degradation of land surfaces there led to a threefold increase in atmospheric dust at that time. It is thus possible for human-induced desertification to generate dust which in turn increases the degree of desertification by reducing rainfall levels.

Dust storms, generated by deflation from land surfaces with limited vegetation cover, occur frequently in the world's drylands. They happen naturally when strong winds attack dry and unvegetated sandy and silty surfaces. Their frequency also varies from year to year in response to fluctuations in rainfall and wind conditions. At present, however, in some parts of the world, the dust entering the atmosphere as a result of dust storms is increasing because of the effects of human activity. In particular, processes such as overgrazing, which are part of the phenomenon of desertification (see part II, section 3), strip the protective vegetation cover from the soil's surface. Elsewhere, surfaces may be rendered more susceptible to wind attack because of ploughing or disturbance by wheeled vehicles.

Atmospheric aerosols can be an important source of cloud-condensation nuclei. Over the world's oceans a major source of such aerosols is **dimethylsulphide** (DMS). This compound is produced by planktonic algae in seawater and then oxidizes in the atmosphere to form sulphate aerosols. Because the albedo of clouds (and thus the Earth's **radiation budget**) is sensitive to the density of cloud-condensation nuclei, any factor that has an impact on planktonic algae may also have an important impact on climate. The production of such plankton could be affected by water pollution in coastal seas or by global warming. Charlson et al. (1992) believe that anthropogenically derived sulphate aerosols could significantly increase planetary albedo, through their direct scattering of short-wavelength solar radiation and their modification of the short-wave reflective properties of clouds. Thus they could exert a cooling influence on the planet. Charlson et al. maintain that this effect could be as great as the current human-induced global warming; but acting, of course, in the opposite way, as 'global cooling'.

A nuclear conflict could produce the most catastrophic effects of anthropogenic aerosols in the atmosphere. Explosion, fire and wind might generate a great pall of smoke and dust in the atmosphere which would make the world dark and cold. It has been estimated that if the exchange reached a level of several thousand megatons, a 'nuclear winter' would occur in which temperatures over much of the world would be depressed to well below freezing point.

Fears were also expressed that the heavy smoke palls generated by oil-well fires in the Gulf War of 1991 might have serious climatic impacts. The actual effects are still not clear. However, preliminary studies have suggested that because most of the smoke generated by the oil-well fires stayed in the lower **troposphere** and remained in the air for only a short time, the effects (some cooling) were local rather than global. It also seems that the operation of the South Asian monsoon was not significantly affected.

Although some of this discussion of the effects of aerosols in the atmosphere is speculative at the global scale, this is not so at the more local scale, where it is clear that human actions can change levels of visibility. This is especially true in urban areas, where the concentration of light-scattering and light-absorbing aerosols in the atmosphere is greatest. For example, before the Clean Air Acts (most notably those passed in 1956 and 1968) London suffered some severe smogs that reduced visibility to a few metres and killed thousands of people (e.g. in the winter of 1952). Reduced burning of coal since the Clean Air Acts has cut down smoke emissions, improving visibility in many parts of Britain. Fogs have become much rarer over the last three decades.

Sulphate emissions from coal-fired power stations have also been reduced. An analysis of changes in visibility at a large number of sites in the UK shows that between 1962 and 1990 the median atmospheric visibility has improved from 10.9 km to 26.0 km (Lee, 1994). Figure III.1 shows the number of days per year when fog occurred in Britain over the period 1950–83. It is clear that although the frequency of fogs has not changed a great deal in coastal areas (where they are largely a natural phenomenon), in the inland industrial heartland they have declined very substantially as a result of Clean Air legislation and changes in industrial technology.

The total suspended particulates (TSP) is the total mass of aerosol particles per volume of air (usually measured in μg per cu metre of air). Of this TSP, much recent concern has focused on the respirable suspended particulates (RSP), particles with diameters of less than 10 μm (also known as PM10s). These small particles are the only ones which can be deposited in the respiratory system – lungs and bronchial tubes – as larger particles are filtered out by the nose, mouth and throat. In many urban areas, concentrations of RSP have become worryingly high. The build-up of particles in lungs can contribute to bronchitis and other respiratory diseases.

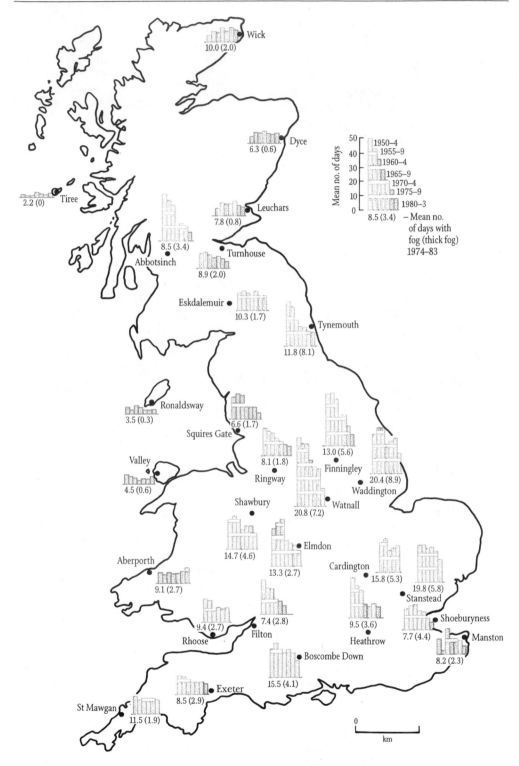

Figure III.1 The spatial variation of fog over Great Britain, 1950–1983
Source: After Musk (1991), fig. 6.6.

The dust bowl

The dust bowl of the 1930s in the Great Plains of the USA is perhaps the best known and most often quoted example of large-scale wind erosion and dust storm activity anywhere in the world. The most severe storms ('black blizzards') occurred in the 'dust bowl' between 1933 and 1938, and were most frequent during the spring of these years. At Amarillo, Texas, at the height of the period, one month had 23 days with at least 10 hours of airborne dust, and in one in five storms visibility was zero. For comparison, the long-term average for this part of Texas is just six dust storms a year.

The reasons for this most dramatic of ecological disasters have been widely discussed. Blame has largely been laid at the feet of the pioneering farmers and 'sod busters' who ploughed up the plains for cultivation. For although dust storms are frequent in the area during dry years, and the 1930s was a drought period, the scale and extent of the 1930s events were unprecedented.

Plate III.1 In the 1930s (the 'dirty thirties') the Great Plains of the USA experienced many 'black blizzards' (dust storms) caused by a combination of a run of dry, hot years, and the ploughing up of large tracts of land for grain production. Similar phenomena occur at the present day in the Sahel zone of West Africa. This example occurred in Mali in 1977. (Rod McIntosh)

Further reading

Goudie, A. S. and Middleton, N. J., 1992, The changing frequency of dust storms through time. *Climatic Change* 20, 197–225.

The Gulf War oil fires: hype and reality

Following the Iraqi invasion of Kuwait on 2 August 1990, deliberate oil spills and oil-well fires were used by the Iraqi leader Saddam Hussein as a weapon of war. In January 1991 Iraqi forces detonated over 800 oil wells (out of a toal of around 1,116 wells in Kuwait), of which 730 exploded. Most of these (656) burned for several months and the remainder gushed out oil. Around 1 billion barrels of crude oil were lost, representing 1.5–2 per cent of the entire Kuwaiti oil reserve.

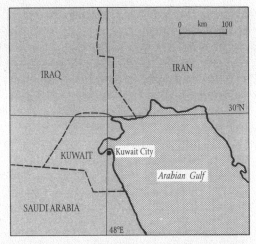

Immediately after this episode scientists and environmental activists speculated that the fires, let alone the spilled oil, would have serious local, regional and global climatic impacts. Doomsday scenarios were suggested including dramatic global cooling, similar to the nuclear winter hypothesis; super-acid rain; diversion of the Asian summer monsoon; and rapid snow melt from falls of black snow.

Later, however, scientific studies involving remote sensing, ground-level monitoring and computer modelling studies showed that the global climatic impacts had been exaggerated. The smoke was not injected high enough to spread over large areas of the Northern Hemisphere; most of it was confined to an altitude of between 1 km and 3 km. Beneath the plume of smoke, daylight and daytime temperatures were reduced. Simulation models suggested a decrease in surface daytime air temperatures of between 4°C and 10°C (Bakan et al., 1991; Browning et al., 1991). There has been no permanent winter, no major diversion of the monsoon, and no super-acid rain.

Scientific studies have shown, however, that the months of burning produced emissions of sulphur dioxide, carbon monoxide, hydrogen sulphide, carbon dioxide and nitrogen oxides (estimates are shown in table III.1). Particulates containing partly burned hydrocarbons and metals such as vanadium and nickel were also discharged into the atmosphere. These emissions may have severe local impacts. For example, monitoring of inhalable particulate matter (PM10s) in the Eastern Province of Saudi Arabia during and after the Kuwaiti oil fires found high concentrations at various places, higher than the maximum permissible level of 340 µg per cu metre (Husain and Amin, 1994). Other studies in Kuwait itself in late April to early May 1991 showed high levels of total airborne particulate matter (soot, organic carbon, sulphate and chloride) but rather low levels of sulphur dioxide, nitrogen dioxide and carbon monoxide.

The local health and ecological impacts of such elevated pollution levels are now of major concern. Some of the compounds released may be carcinogenic. The inhalable PM10s may cause severe health problems. Hospital studies in Kuwait in 1991 showed a moderate rise (about 6 per cent) in lung and heart complaints (Hoffman, 1991). Clearly, long-term health issues need monitoring.

Plate III.2   In the Gulf War of 1991, large quantities of oil were burnt as here at the Al Burgan oil field. Fears were expressed that this could have a severe climatic impact. In the event these fears were to a large extent misplaced. (E.P.L./Jim Hodson)

Table III.1   Predicted annual production of Kuwaiti oil fires in 1991

Type of emission	Amount (Tg per year)[a]	Comparison with current global emissions
Fine particulate black smoke	5	Roughly one-third of carbon particles produced by tropical biomass burning
Sulphur (as S oxides)	2	Slightly more than current UK annual S emissions
Nitrogen (as N oxides)	0.5	1988 UK emissions of nitrogen oxide were 0.75 Tg
Carbon (ultimately as CO_2)	60	About 1% of current global annual CO_2 emissions from fossil fuel combustion

[a] $Tg = Teragramme = 1 \times 10^{12}g$.
Source: Browning et al. (1991).

3   ANTHROPOGENIC CLIMATE CHANGE: THE ROLE OF LAND COVER CHANGES

Another major possible human-induced cause of climate change is change in the reflectivity (albedo) of the ground surface and the proportion of solar radiation which the surface reflects. Land-use changes create differences in albedo which have important effects on the energy balance of an area. Tall rain forest may have an albedo as low as 9 per cent, while the albedo of a desert may be as high as 37 per cent. There has been growing interest recently in the possible consequences of deforestation on climate as a result of the associated change in albedo. Ground deprived of vegetation cover as a result of deforestation and overgrazing (as in parts of the Sahel) has a very much higher albedo than ground covered in plants. This could affect temperature levels. Satellite imagery of the Sinai–Negev region of the Middle East shows an enormous difference in image between the relatively dark Negev and the very bright Sinai–Gaza Strip area. This line coincides with the 1948–9 armistice line between Israel and Egypt and results from different land-use and population pressures on either side of that boundary. Otterman (1974) has suggested that the albedo affected by land use has produced temperature changes of the order of 5°C.

Charney et al. (1975) have argued that the increase in surface albedo resulting from a decrease in plant cover would lead to a reflection outwards of incoming radiation, and an increase in the radiative cooling of the air. Consequently, they argue, the air would sink to maintain thermal equilibrium by **adiabatic compression**, and cumulus convection and its associated rainfall would be suppressed. A positive feedback mechanism would appear at this stage: namely, the lower rainfall would in turn adversely affect plants and lead to a further decrease in plant cover.

This view was disputed by Ripley (1976). He suggested that Charney and his colleagues, when considering the impact of vegetation changes on albedo, failed to consider the effect of vegetation on evapotranspiration. He pointed out that vegetated surfaces are usually cooler than bare ground, since much of the solar energy absorbed is used to evaporate water. He concluded from this that protection from overgrazing and deforestation might, in contrast to Charney's views, be expected to lower surface temperatures and thereby reduce, rather than increase, convection and precipitation.

The models used by some scholars suggest that removal of the humid tropical rain forests could also have direct climatic effects. Lean and Warrilow (1989) used a **general circulation model** (GCM) which suggested that deforestation in the Amazon basin would lead to reductions in both precipitation and evaporation as a result of the changes in surface roughness and albedo. The surface roughness effect occurs because rain forest has quite a jagged canopy, and this in turn affects wind flow. Likewise, a UK Meteorological Office GCM shows that the deforestation of both Amazonia and Zaire would cause precipitation levels to fall by changing surface albedo (Mylne and Rowntree, 1992).

Budyko (1974) believes that the present use of irrigation over about 0.4 per cent of the Earth's surface (1.3 per cent of the land surface) is decreasing the albedo of irrigated areas, possibly on average by 10 per cent. The corresponding change in the albedo of the entire Earth–atmosphere system would amount to about 0.03 per cent; enough, according to Budyko, to maintain the global mean temperature at a level nearly 0.1°C higher than it would otherwise be.

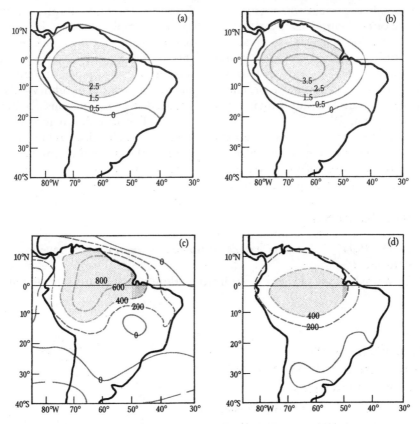

Figure III.2   Predictions of the change in climate following a conversion of Amazonian rain forest to grassland. (a) Temperature increase (°C); (b) Evaporation decrease (mm per year); (c) Rainfall decrease (mm per year); (d) Evapotranspiration decrease (mm per year)
Source: After Shukla et al. (1990).

A change in land use can also lead to a change in the moisture content of the atmosphere. It is possible, for example, that if humid tropical rain forests are cut down, the amount of moisture transpired into the atmosphere above them will be reduced. This would reduce the potential for rain (figure III.2(c)). The spread of irrigation could have the opposite effect, leading to increased atmospheric humidity levels in the world's drylands. The High Plains of the USA, for example, are normally covered with sparse grasses and have dry soils throughout the summer. Evapo-

transpiration there is very low. In the last four decades, however, irrigation has been developed throughout large parts of the area. This has greatly increased summer evapotranspiration levels. There is strong statistical evidence that rainfall in the warm season has been increased by the use of irrigation in two parts of this area: one extending through Kansas, Nebraska and Colorado, and a second in the Texas Panhandle. The largest absolute increase was in the latter area. Significantly, it occurred in June, the wettest of the three heavily irrigated months. The effect appears to be especially important when stationary **weather fronts** occur. This is a situation which allows for maximum interaction between the damp irrigated surface and the atmosphere. Hail storms and tornadoes are also significantly more prevalent over irrigated than over non-irrigated regions (Nicholson, 1988).

Although we have discussed albedo change and atmospheric moisture changes as two separate classes of processes, they need to be seen as working together, and also in association with other mechanisms. For an example of why this is important, we can look at tropical rain-forest removal. This causes albedo change, reduction in moisture loss by evapotranspiration, and a change in surface roughness. The combined effects may be considerable (figure III.2). They include an increase in temperature, a major decrease in loss of moisture into the atmosphere, and a very major decrease in rainfall.

FURTHER READING

Kemp, D. D., 1994, *Global Environmental Issues: A Climatological Approach*, 2nd edn. London: Routledge.
A well-illustrated, clear and accessible introduction to many areas of global climatic change.

4   THE ENHANCED GREENHOUSE EFFECT AND GLOBAL WARMING

Planet Earth receives warmth from the sun. Radiation from the sun is partly trapped by the atmosphere. It passes through the atmosphere and heats the Earth's surface. The warmed surface radiates energy, but at a longer wavelength than sunshine. Some of this energy is absorbed by the atmosphere, which as a result warms up. The rest of the energy escapes to space. We call this process of warming the 'greenhouse effect' because the atmosphere is perceived to act rather like glass in a greenhouse (figure III.3). Although the atmosphere consists primarily of nitrogen and oxygen, it is some of the so-called **trace gases** which absorb most of the heat, in spite of the fact that they occur in very small concentrations. These are called the 'greenhouse gases'.

Various greenhouse gases occur naturally – water vapour (H_2O), carbon dioxide (CO_2), methane (CH_4), ozone (O_3)

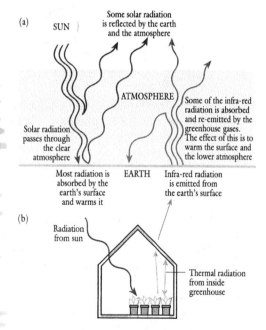

Figure III.3   (a) The greenhouse effect in the atmosphere. (b) A diagram showing how a greenhouse acts as a 'radiation blanket'.
Sources: (a) Houghton et al. (1990), figure 1; (b) Houghton (1994), figure 2.2.

and nitrous oxide (N_2O). In recent centuries and decades, however, the quantities of some of these greenhouse gases have started to increase because of human activities. In addition, a new type of greenhouse gas, the chloroflurocarbons (CFCs), has been introduced to the atmosphere in the last fifty years.

Since the start of the industrial revolution humans have been taking stored carbon out of the earth in the form of fossil fuels (coal, oil and natural gas). They burn these fuels, releasing CO_2 in the process. The pre-industrial level of CO_2 in the atmosphere may have been as low as 260–270 parts per million by volume (ppmv). The present level exceeds 350 ppmv and is still rising, as is evident in records of atmospheric composition from various parts of the world. Fossil fuel burning and cement manufacture release over 6 gigatonnes of carbon to the atmosphere as CO_2 each year. Burning of forests and changes in the levels of organic carbon in soils subjected to deforestation and cultivation may also contribute substantially to CO_2 levels in the atmosphere, perhaps by around 2 gigatonnes of carbon each year.

Other gases, as well as CO_2, will probably contribute to the accelerated greenhouse effect. The effect of each on its own may be relatively small, but the effects of all of them combined may be considerable. Moreover, molecule for molecule, some of these other gases may be more effective as greenhouse gases than CO_2. This applies to methane (CH_4) which is 21 times more effective than CO_2, to nitrous oxide (N_2O) which is 206 times more effective, and to the CFCs, which are 12,000–16,000 times more effective.

Where do these other gases come from and why are amounts of them increasing? Concentrations of methane are now over 1,600 parts per billion by volume (ppbv), compared to eighteenth-century background levels of 600 ppbv. Methane has increased as a result of the spread of rice cultivation in waterlogged paddy fields, enteric fermentation in the growing numbers of belching and flatulent domestic cattle, and the burning of oil and natural gas. Nitrous oxide levels have increased because of the combustion of hydrocarbon fuels, the use of ammonia-based synthetic fertilizers, deforestation and vegetation burning. The increase in CFCs in the atmosphere (which is also associated with ozone depletion in the stratosphere – see section 7 below) results from their use as refrigerants, as foam makers, as fire control agents, and as propellants in aerosol cans. Use of CFCs is now being restricted by various international agreements.

The Earth's climate has become generally warmer over the last century or so, and the 1980s saw an unprecedented number of warm years. This has prompted some scientists to propose that global warming, as a result of the accelerated greenhouse effect, has already started. However, the complexity of factors that can cause climatic fluctuations leads many scientists to doubt that the case is yet fully proven. Most, however, believe that if concentrations of effective greenhouse gases continue to rise, and attain double their natural levels by around the middle of the twenty-first century, then temperatures will rise by several degrees over that period. The Intergovernmental Panel on Climate Change (IPCC), which reported in 1990, suggested that global mean temperature might increase during the next century at a rate of 0.3°C per decade. The IPCC report of 1996 suggested a 'best estimate' of 2.0°C increase in temperature by 2100 (with a range of 1–3.5°C). This is somewhat lower than previous predictions because of improvements in information and modelling techniques. Cooling effects of aerosols are taken into account in this prediction. The rise in temperature

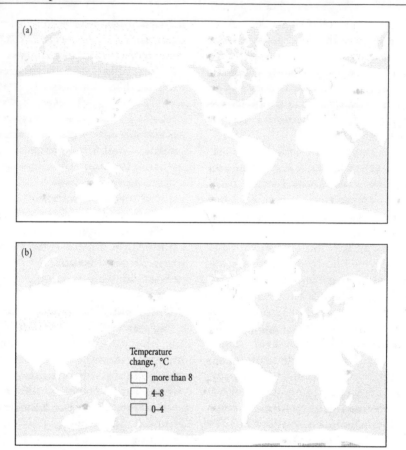

Figure III.4   Change in global surface temperature following a doubling of CO_2: (a) December, January and February; (b) June, July and August
Source: Kemp (1994), figure 7.8, using data in Houghton et al. (1990).

will not, however, be the same across the globe. In particular, high latitudes (e.g. northern Canada and Eurasia) will show even more pronounced warming, perhaps two to three times the global average (figure III.4).

Such increases in temperature, if they occur, will undoubtedly cause major changes in the general atmospheric circulation. These in turn will cause changes in precipitation patterns. Overall levels of precipitation over the globe will increase as more moisture is released by higher rates of evaporation from the oceans. However,

some areas will get wetter while some will get drier. There is still considerable uncertainty about what precise pattern precipitation will take as a result of these changes. The very cold, dry areas of high latitudes may well become moister as a warmer atmosphere will be able to hold more moisture. Some tropical areas may receive more rain as the vigour of the monsoonal circulation and of tropical cyclones is increased. Some mid-latitude areas, like the High Plains of America, may become markedly drier.

There is, however, great uncertainty as

to how far the climate may change as a result of the greenhouse effect. The reasons for this uncertainty include:

- doubts about how fast the global economy will grow;
- doubts about what fuels will be used in the future;
- doubts about the speed at which land-use changes are taking place;
- uncertainty regarding how much CO_2 will be absorbed by the oceans and by biota;
- uncertainties about the role of other anthropogenic and natural (e.g. volcanic) causes of climatic change;
- the assumptions that are built into many of our predictive general circulation models (e.g. about the role of clouds);
- the role of possible positive feedbacks and thresholds that may mean changes are more sudden than anticipated, or do not happen at all.

The degree of global warming that is proposed for the coming decades does not at first sight appear enormous. However, it may, over a period that is very short in geological terms, produce warmer conditions than have existed for several million years and set up a series of changes that have important implications both for the environment and for humans. Some of these implications may be benign (e.g.

warmer conditions will enable new crops to be grown in Britain) but some of them will be malign (e.g. more frequent and longer droughts in the High Plains of America). Among the *possible* environmental consequences are:

- more intense, widespread and frequent tropical cyclones;
- the melting of alpine glaciers;
- the degradation of permafrost in tundra areas;
- the wholesale displacement of major vegetation belts such as the boreal forests of the Northern Hemisphere;
- rising sea levels and associated flooding of coral reefs, deltas, wetlands, etc., and accelerated rates of beach erosion;
- decreased flow of water in streams as a result of increased loss of moisture by evapotranspiration;
- reduction in the extent of sea ice in polar waters;
- shifts in the range of certain vector-borne diseases (e.g. malaria).

Many scientists and politicians believe that the case has now been made that global warming will occur and that the resulting changes are likely to be so significant that action needs to be taken. In some countries a policy of 'no regrets' is being promoted. This is a policy under which the reduction of greenhouse emissions is also justifiable on other grounds (see table III.2).

Table III.2 Examples of 'no regrets' climate-warming policies		
Policy	*Effect on greenhouse gases*	*Other beneficial effects*
Tree planting	Increased biosphere sink strength to absorb CO_2	Improved microclimate
		Improved habitat for many species
		Reduced soil erosion
		Reduced seasonal peak river flows
Energy conservation	Reduced CO_2 emissions	Conservation of non-renewable resources for current and future generations
Energy efficiency	Reduced CO_2 emissions	Conservation of non-renewable resources for current and future generations
CFC emission control	Reduced CFC emissions	Reduced stratospheric ozone-layer depletion (see section 7)
		Reduced surface UV-B and associated skin cancer and blindness

FURTHER READING

Houghton, J. T., 1994, *Global Warming: The Complete Briefing*. Oxford: Lion Books. A useful, clearly written introduction by a leading expert that summarizes the key findings of the world's scientific community in this area.

Houghton, J. T., Jenkins, G. J. and Ephraums, J. J. (eds), 1990, *Climate Change: The IPCC Scientific Assessment*. Cambridge: Cambridge University Press.
Houghton, J. T., Callander, B. A. and Varney, S. K. (eds), 1992, *Climate Change 1992: The Supplementary Report of the IPCC Scientific Assessment*. Cambridge: Cambridge University Press.
Houghton, J. T., Meira Filho, L. G., Callander, B. A., Harris, N., Kaltenberg, A. and Maskell, K. (eds), 1996, *Climate Change 1995: The Science of Climate Change*. Cambridge: Cambridge University Press.
Three reports from the global body, the Intergovernmental Panel on Climate Change (IPCC), established to look at the causes and consequences of global warming.

Kemp, D. D., 1994, *Global Environmental Issues: A Climatological Approach*. London: Routledge.

Global warming and UK agriculture

As a result of global warming the temperature in Britain could rise by several degrees Celsius during the course of the next 50–100 years. A change in the climate of this magnitude would be likely to shift the thermal limits of agriculture by around 300 km of latitude and 200 m of altitude per degree Celsius. Several crop species, such as wheat, maize and sunflowers, have their contemporary northern limits in the UK. An increase of temperature could, therefore, assuming that soil conditions were suitable, lead to a substantial northward shift of cropping zones. This could transform the British agricultural landscape. British fields and rural areas might come to resemble those currently found further south in mainland Europe. For example, the northern limit of grain maize, which currently lies in the extreme south of England (see figure III.5), could be shifted across central England by a 0.5°C increase in temperature, across northern England by a 1.5°C increase and into the north of Scotland by an increase of 3°C.

A rise in temperature, apart from transforming the range over which particular crop types could be grown, could be significant for the agricultural sector in other ways. For example, higher temperatures and more frequent summer droughts might reduce crop yields. The occurrence of certain plant pests and diseases could change, for better or worse.

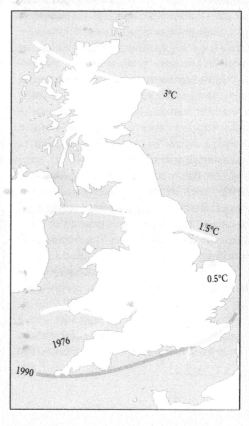

Figure III.5 The potential distribution of grain maize in the UK under different warming scenarios
Source: After Parry in Jones (1993), fig. 11.

Further reading

Jones, D. K. C. (ed.), 1993, Earth surface resources management in a warmer Britain. *Geographical Journal* 159, 124–208.

5   URBAN CLIMATES

Climate statistics for recent decades show that many cities have become warmer than the countryside around them. Climatologists have long spoken of the urban 'heat island' in the cool rural 'sea'. The boundary between countryside and city forms a steep temperature gradient or 'cliff' to the urban heat island. Much of the rest of the urban area appears as a 'plateau' of warm air with a steady but shallower gradient of increasing warmth towards the city centre. The urban core, or central business district, with its high-density buildings, is a 'peak' where the maximum temperature is found. The difference between this peak value and that in the rural 'sea' defines the intensity of the urban heat island.

There are various reasons why cities may be relatively warmer than the rural areas that surround them (figure III.6). First, city surfaces absorb significantly more radiation from the sun than rural surfaces. This is because a higher proportion of the reflected radiation is retained by the high walls and dark-coloured roofs and roads of the city environment. These city surfaces have both great thermal capacity and high conductivity, so that heat is stored during the day and released by night. By contrast, vegetation cover gives plant-covered rural areas an insulating 'blanket', so that they experience relatively lower temperatures both by day and by night. This effect is enhanced and compounded by the evaporation and transpiration that occur from plant-covered surfaces. Secondly, cities are relatively warm because they generate a large amount of artificial heat. Energy is produced and then used by industrial, commercial, transport and domestic users.

The heat island effect is not the only way that towns and cities affect the climate. However, the effects of urban areas on other aspects of climate are less easily measured and explained. There is some evidence that rainfall, including that produced by summer thunderstorms, can be

Figure III.6   Mechanisms of urban climates

higher over urban than rural areas. There are various possible reasons for this:

- the urban heat island generates convection (i.e. thermally induced upward movement of air);
- the presence of high-rise buildings and a mixture of building heights induces air turbulence and promotes increased vertical motion;
- cities may produce large amounts of water vapour from industrial sources and power stations, and also various pollutant aerosols that act as condensation nuclei.

The London area provides an interesting but by no means unique example of the effects of large cities on precipitation levels. In this case it seems that the mechanical effect of the city was the main cause of local peak precipitation. It had this effect by being a mechanical obstacle to air flow, on the one hand, and by causing frictional convergence of flow, on the other. A long-term analysis of thunderstorm records for south-east England shows that thunderstorms are more frequent over the urban area than elsewhere in the region (Atkinson, 1968). The similarity between the shape of the thunderstorm isopleth and that of the urban area is striking. Moreover, Brimblecombe (1977) found that thunderstorms have become steadily more frequent as the city has grown.

Winds are another aspect of the urban climate. There are two main aspects to the effect that cities have on winds: first, the rougher surface cities present in comparison with rural areas; and secondly, the frequently higher temperatures of the city. Buildings, especially those in cities with a very varied skyline, exert a powerful frictional drag on air moving over and around them. This creates turbulence, with rapid and abrupt changes in both direction and speed. The average speed of the winds is lower in built-up areas than over rural areas. However, Chandler (1976) found that in London, when winds are light, speeds are greater in the inner city than outside, whereas when winds are strong speeds are greater outside the city centre and lower within it. Overall, annual wind speed in central London is about 6 per cent lower than outside, but for the higher-velocity winds (more than 1.5 metres per second) the reduction is more than twice that.

Studies in two English cities, Leicester and London, have shown that on calm, clear nights, when the urban heat island effect is at its greatest, there is a surface inflow of cool air towards the warmest zones. These so-called 'country breezes' are low in velocity and are quickly slowed down further by intense surface friction in the suburban areas. One effect of these breezes is to transport pollution from the outer parts of an urban area into the city centre, accentuating the pollution problem during periods with photochemical smogs.

FURTHER READING

Landsberg, H. E., 1981, *The Urban Climate*. New York: Academic Press.
The classic study.

Oke, T. J., 1987, *Boundary Layer Climates*, 2nd edn. London: Routledge.
A thorough review of local-scale climates which includes an authoritative study of urban climates.

The implications of some urban heat islands

As cities grow, so does their heat island effect. In Columbia, Maryland, USA, for example, when the town had only 1,000 inhabitants in 1968, the maximum temperature difference between residential areas and the surrounding countryside was only 1°C. By 1974, when it had grown to a town with a little over 20,000 inhabitants, the maximum heat island effect had grown to 7°C.

Thus the annual average temperatures over the hearts of great cities can be substantially higher than those over the surrounding countryside. This is clear from the temperature map of Paris (figure III.7(a)). The outlying weather stations have mean annual temperatures of 10.6–10.9°C, whereas in central Paris the value is 12.3°C, about 1.5°C higher. These values have all been reduced to a uniform elevation of 50 metres above sea level to correct for possible **orographic** effects.

Urban climates are often characterized by different precipitation characteristics from rural areas. For example, it is remarkable that there tends to be more rain in Paris during the week than at weekends (figure III.7(b)). There is a gradual increase in average rainfall from Monday to Friday (when factories are producing more heat and aerosols), then a sharp drop for Saturdays and Sundays. The weekend average for May to October was 1.47 mm, whereas the workday average was 1.93 mm – a decrease of 24 per cent for the weekend.

In winter months the consequences of urban heat islands can be particularly significant in cold regions. For example, the average date of the last freezing temperatures at the end of winter in Washington DC in the USA is about three weeks earlier than in the surrounding rural areas (figure III.7(c)). In autumn the city has, on average, the first freezing temperature on about 3 November, whereas in the outlying suburbs 0°C will usually be observed about two weeks earlier. Thus in all the frost-free season will be about 35 days longer than it is in the countryside. Similar figures have been obtained for some other great cities. Data for Moscow, Russia, suggest an increase of around 30 days without freezing, while those for Munich in Germany suggest an increase that can be as great as 61 days.

In summer months, the urban heat island effect can lead to an increasing demand for air conditioning, and because the energy requirements of air conditioning are greater than those of heating, the savings in winter heating bills are more than offset. Moreover, air conditioning can aggravate the heat island effect, because air conditioning plant discharges heat to the outside air, where it mixes with air that has already been warmed up by the hot air forming adjacent to sunlit walls and pavements.

Further reading

Landsberg, H. E., 1981, *The Urban Climate*. New York: Academic Press.

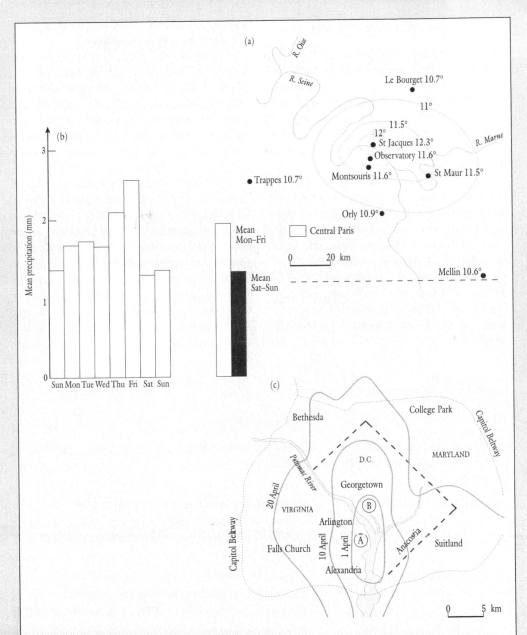

Figure III.7 The impact of urban areas on climate. (a) Annual isotherms in the Paris region; (b) Precipitation in Paris, averaged by day of the week; (c) Average date of last freezing temperature in spring in Washington, DC (A = International Airport; B = White House)
Source: Landsberg (1981), figures 5.5, 8.6, 5.25.

ӠAN AIR POLLUTION

entration of large numbers of people, factories, power stations and cars mean that large amounts of pollutants may be emitted into urban atmospheres. If weather conditions permit, the level of pollution may build up. The nature of the pollutants (table III.3) has changed as technologies have changed. For example, in the early phases of the industrial revolution in Britain the prime cause of air pollution in cities may have been the burning of coal, whereas now it may be vehicular emissions. Different cities may have very different levels of pollution, depending on factors such as the level of technology, size, wealth and anti-pollution legislation. Differences may also arise because of local topographic and climatic conditions. Photochemical smogs, for example, are a more serious threat in areas subjected to intense sunlight.

The variations in pollution levels between different cities are brought out in figure III.8, which shows data for two types of pollution for a large range of city types. The data were prepared for the years 1980–4 by the Global Environment Monitoring System of the United Nations Environment Programme (UNEP). Figure III.8(a) shows concentrations of total particulate matter. Most of this comes from the burning of poor-quality fuels. The shaded horizontal bar indicates the range of concentrations that UNEP considers a reasonable target for preserving human health. Note that the annual mean levels range from a low of about 35 μg per cu metre to a high of about 800 μg per cu metre: a range of about 25-fold! The higher values appear to be for rapidly

Table III.3 Major urban pollutants

Type	Some consequences
Suspended particulate matter (characteristically 0.1–25 μm in diameter)	Fog, respiratory problems, carcinogens, soiling of buildings.
Sulphur dioxide (SO_2)	Respiratory problems, can cause asthma attacks. Damage to plants and lichens, corrosion of buildings and materials, production of haze and acid rain.
Photochemical oxidants: ozone and peroxyacetyl nitrate (PAN)	Headaches, eye irritation, coughs, chest discomfort, damage to materials (e.g. rubber), damage to crops and natural vegetation, smog.
Oxides of nitrogen (NOx)	Photochemical reactions, accelerated weathering of buildings, respiratory problems, production of acid rain and haze.
Carbon monoxide (CO)	Heart problems, headaches, fatigue, etc.
Toxic metals: lead	Poisoning, reduced educational attainments and increased behavioural difficulties in children.
Toxic chemicals: dioxins, etc.	Poisoning, cancers, etc.

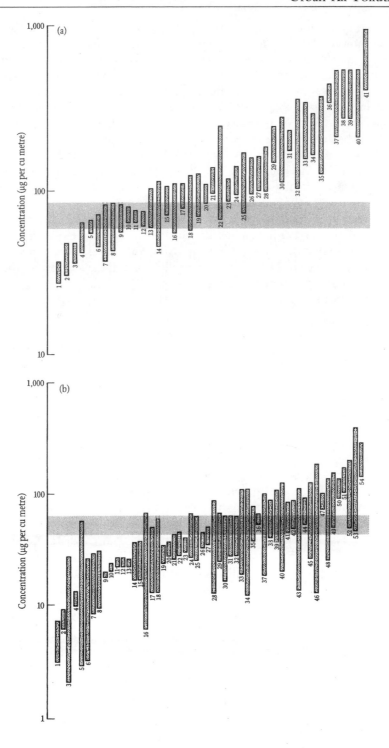

(For caption see overleaf)

growing cities in the developing countries. Some cities, however, such as Kuwait, may have unusually high values because of their susceptibility to dust storms from desert hinterlands. The lower values tend to come from cities in developed areas (e.g. Western Europe, Japan and North America).

Figure III.8(b) shows concentrations for sulphur dioxide. Much of this gas probably comes from the burning of high-sulphur coal. Once again, the horizontal shaded bar indicates the concentration range considered by UNEP to be a reasonable target for preserving human health. These data indicate that the concentrations of sulphur dioxide can differ by as much as three times among different sites within the same urban area and by as much as 30 times between different urban areas.

In some cities concentrations of pollutants have tended to fall over recent decades. This can result from changes in industrial technology or from legislative changes (e.g. clean air legislation, restrictions on car use, etc.). In many British cities, for example, legislation since the 1950s has reduced the burning of coal. As a consequence, fogs have become less frequent and the amount of sunshine has increased. Figure III.9 shows the overall trends for the United Kingdom, and highlights the decreasing fog frequency and increasing sunshine levels. The concentrations of various pollutants have also been reduced in the Los Angeles area of California (figure III.10). Here, carbon monoxide, non-methane hydrocarbon, nitrogen oxide and ozone concentrations have all fallen steadily over the period since the late 1960s.

However, both of these examples of improving trends come from developed countries. In many cities in poorer countries, pollution is increasing at present. In certain countries, heavy reliance on coal, oil and even wood for domestic cooking

Figure III.8   (a) The range of annual averages of total particulate matter concentrations measured at multiple sites within 41 cities, 1980–1984. Each numbered bar represents a city, as follows: 1, Frankfurt; 2, Copenhagen; 3, Cali; 4, Osaka; 5, Tokyo; 6, New York; 7, Vancouver; 8, Montreal; 9, Fairfield; 10, Chattanooga; 11, Medellin; 12, Melbourne; 13, Toronto; 14, Craiova; 15, Houston; 16, Sydney; 17, Hamilton; 18, Helsinki; 19, Birmingham; 20, Caracas; 21, Chicago; 22, Manila; 23, Lisbon; 24, Accra; 25, Bucharest; 26, Rio de Janeiro; 27, Zagreb; 28, Kuala Lumpur; 29, Bombay; 30, Bangkok; 31, Illigan City; 32, Guangzhou; 33, Shanghai; 34, Jakarta; 35, Tehran; 36, Calcutta; 37, Beijing; 38, New Delhi; 39, Xian; 40, Shenyang; 41, Kuwait City. (b) The range of annual averages of sulphur dioxide concentrations measured at multiple sites within 54 cities, 1980–1984. Each numbered bar represents a city, as follows: 1, Craiova; 2, Melbourne; 3, Auckland; 4, Cali; 5, Tel Aviv; 6, Bucharest; 7, Vancouver; 8, Toronto; 9, Bangkok; 10, Chicago; 11, Houston; 12, Kuala Lumpur; 13, Munich; 14, Helsinki; 15, Lisbon; 16, Sydney; 17, Christchurch; 18, Bombay; 19, Copenhagen; 20, Amsterdam; 21, Hamilton; 22, Osaka; 23, Caracas; 24, Tokyo; 25, Wroclaw; 26, Athens; 27, Warsaw; 28, New Delhi; 29, Montreal; 30, Medellin; 31, St Louis; 32, Dublin; 33, Hong Kong; 34, Shanghai; 35, New York; 36, London; 37, Calcutta; 38, Brussels; 39, Santiago; 40, Zagreb; 41, Frankfurt; 42, Glasgow; 43, Guangzhou; 44, Manila; 45, Madrid; 46, Beijing; 47, Paris; 48, Xi'an; 49, São Paulo; 50, Rio de Janeiro; 51, Seoul; 52, Tehran; 53, Shenyang; 54, Milan *Source*: Graedel and Crutzen (1993).

and heating means that their levels of sulphur dioxide and suspended particulate matter (SPM) are high and climbing. In addition, rapid economic development is bringing increased emissions from industry and motor vehicles, which are generating progressively more serious air-quality problems.

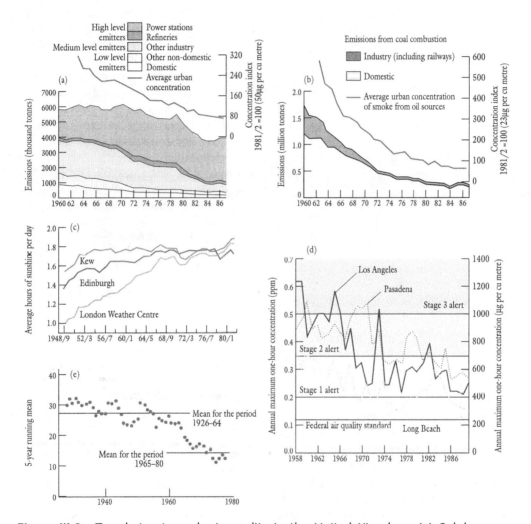

Figure III.9 Trends in atmospheric quality in the United Kingdom: (a) Sulphur dioxide emissions from fuel combustion and average urban concentrations; (b) Smoke emissions from coal combustion and average urban concentrations of oil smoke; (c) Increase in winter sunshine (10-year moving average) for London and Edinburgh city centres and for Kew, outer London; (d) Annual maximum hourly ozone concentrations at selected sites in the Los Angeles basin, 1958–1989; (e) Annual fog frequency at 0900 GMT in Oxford, central England, 1926–1980

Sources: (a)–(c) Figures from Department of Environment; (d) After Elsom (1992), figure 2.11; (e) After Gomez and Smith (1984), figure 3.

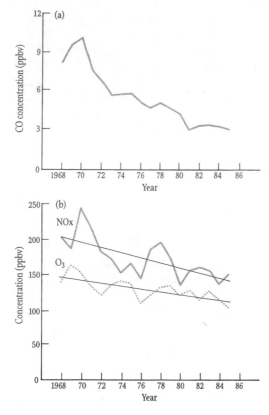

Figure III.10   Air quality trends in Los Angeles and its environs have been measured continuously and averaged over each hour. The highest of the hourly averages is then selected for trend analysis. Part (a) shows the downward trend in carbon monoxide (CO) concentrations; this trend is consistent with vehicular emission control measures; part (b) shows the trend for oxides of nitrogen (NOx) and ozone (O$_3$). Both are expressed in parts per billion by volume (ppbv).
Source: Modified from Kuntesal and Chang (1987), copyright 1987 by Air Pollution Control Association.

Particular attention is being paid at the present time to the chemical composition of SPMs, and particularly to those particles that are small enough to be breathed in (i.e. smaller than 10 µm, and so often known as PM10s). Also of great concern in terms of human health are elemental carbon (for example, from diesel vehicles), polynuclear aromatic hydrocarbons (PAHs) and toxic base metals (e.g. arsenic, lead, cadmium and mercury), in part because of their possible role as **carcinogens**.

Urban air pollution was particularly severe in the former Communist states of Eastern Europe. Carter and Turnock (1993, p. 63) described this problem, and its political background, in the context of Czechoslovakia (now the Czech Republic and Slovakia):

Environmental quality has clearly deteriorated as a result of human activity; the major cause is an excessive and inconsiderate extraction of natural resources, extensive waste emissions and failure to observe ecological and aesthetic laws. These were compounded by an inefficient economy, which consumed inordinate amounts of raw materials and energy, based on outmoded technology which produced manufactured goods with little respect for the ecological consequences. This sad situation was further aggravated by inadequate financial resource allocation for environmental protection, which was of a remedial character rather than one of damage prevention. Much of the blame for this state of affairs must be laid upon the Communist government over the past forty years when legislative, executive and political power was concentrated in the hands of a small controlling group (*nomenklatura*) who did little to correct adverse effects on the environment caused by their policies. Added to this detrimental domestic attitude was the significant contribution made by transboundary pollution from neighbouring states, particularly along the northern and western boundaries of the country.

The problem was exacerbated by the use of lignite (brown coal) in some of the East

Plate III.3   Some of the worst urban air pollution occurred in the former Soviet Union. A particularly gruesome pollution hot spot was the Magnitogorsk steel-making area. (Katz Pictures)

European states. Such coal is of low quality, so disproportionately large quantities have to be burnt; it can also have a very high sulphur content. Carter and Turnock (1993, p. 189) refer to the 'deadly pall of sulphurous smoke' that this fuel source has helped to promote. They point out that even in the late 1980s over three-quarters of Poland's energy came from brown coal, as did two-thirds of the energy in Czechoslovakia and the former East Germany.

Further Reading

Brimblecombe, P., 1987, *The Big Smoke*. London: Methuen.
A diverting history of air pollution, with particular reference to London.

Carter, F. W. and Turnock, D. (eds), 1993, *Environmental Problems in Eastern Europe*. London: Routledge.
An edited collection of papers on the legacy of dreadful air pollution problems in Eastern Europe.

Elsom, D., 1996, *Smog Alert: Managing Urban Air Quality*. London: Earthscan.
A very readable and informative guide.

Air pollution in South African cities: the legacy of apartheid

South Africa produces the world's cheapest electricity, but for many years it has only been available to around 30 per cent of the population. This bald statement sums up the major causes of South African urban air pollution. Around 83 per cent of South Africa's electricity is generated by coal-fired power stations, which burn coal with a sulphur content of around 1.2 per cent. Such high sulphur content (relative to many other types of coal) produces high levels of polluting gases. Many of these plants are located in the eastern Transvaal, which suffers

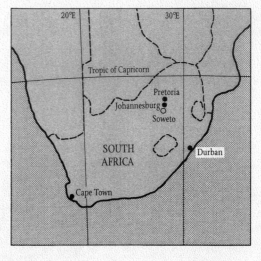

greatly from air pollution and acid deposition. Apartheid, the system of government which dominated South Africa from 1948 to 1994 and forced different racial groups to live apart, produced highly unequal distributions of access to energy resources and of pollution conditions. Apartheid forced black and Coloured populations into poor 'townships', usually without electricity and with severe pollution problems.

The background air pollution in many parts of South Africa is increased in urban environments, where coal, paraffin and wood are burnt as important domestic energy sources. By 1985 about 57 per cent of the entire South African population lived in cities, many of them in squatter settlements and 'townships' where electricity supply was limited. Soweto, for example, some 15 km from Johannesburg, covers nearly 60,000 sq km and had a population in 1990 of around 2.5 million according to some estimates. Electricity was brought into Soweto in 1981, but coal is still widely used as it is cheaper and the supply more reliable.

Sulphur dioxide pollution is now a critical health problem for Sowetan inhabitants. In Soweto, mean annual sulphur concentrations are up to 60 µg per cu metre, whereas in the unpolluted north-west of Transvaal mean annual concentrations are on average only about 7 µg per cu metre. There are also high levels of nitrogen oxides and carbon monoxide in Soweto.

The air pollution problems of Soweto are exacerbated by climate and topography. North and north-westerly winds transport pollution here from Johannesburg city centre, and winter temperature inversions help trap the pollution. The South African Department of Health now believes coal smoke in townships to be the most serious national air pollution problem. Air pollution is a problem indoors as well as outdoors, especially in areas where open fires or inefficient stoves are used for cooking. Suspended particulate matter, carbon

Plate III.4   Air pollution in Cape Town, South Africa. Much of the pollution is caused by the burning of low-quality fuel in the densely populated townships that surround the city. At some times of year the polluting smoke and gases are trapped by climatic conditions called 'inversions'. (A. S. Goudie)

monoxide, oxides of sulphur and nitrogen, hydrocarbons and a range of other pollutants are produced by stoves and fires. The accumulation of indoor and outdoor pollution in Soweto and many other towns is leading to severe respiratory problems, especially in the poorest and most vulnerable members of society. Asbestos also poses an air pollution problem in South Africa, where blue asbestos is mined in the northern Transvaal and northern Cape. Asbestos can cause lung and other cancers, and urbanized areas near mining operations are particularly vulnerable to wind-blown asbestos.

Further reading

Ramphele, M., 1991, *Restoring the Land: Environment and Change in Post-Apartheid South Africa*. London: Panos.

Vogel, C. H. and Drummond, J. H., 1995, Shades of 'green' and 'brown': environmental issues in South Africa. In A. Lemon (ed.), *The Geography of Change in South Africa*, 85–98. Chichester: Wiley.

7   Ozone Depletion and Ozone Pollution

Ozone (O_3) was discovered in 1840. It is a naturally occurring form of oxygen which consists of three oxygen atoms rather than two. It exists throughout the atmosphere in very low concentrations, never exceeding around one molecule in every 100,000 present. It is especially abundant in the stratosphere between 10 and 40 km above the ground. This 'ozone layer' contains about 90 per cent of atmospheric ozone and is important because it provides a thin veil which absorbs ultraviolet (**UV**) radiation from the sun. Indeed, the ozone layer prevents about 97 per cent of UV-B light from reaching the Earth's surface. Too much ultraviolet radiation can damage plants, including the **phytoplankton** that live in the oceans. In humans, it can cause skin cancers; it may also cause eye cataracts and damage the body's immune system. Thus it is clear that any reduction in the thickness and concentration of ozone in the ozone layer is worrying.

In the 1980s satellite observations, ground measurements, and readings from instruments on balloons and in aircraft began to suggest that the ozone layer was becoming thinner, especially over the Antarctic. More recent measurements have indicated that the ozone layer is also thinning over America and northern Europe (see table III.4). Here, ozone decreased on average by around 3 per cent in the 1980s. In the 1970s concern was expressed about possible damage to the ozone layer by high-flying supersonic aircraft such as military jets or Concorde. However, current concern among scientists is focused on a range of manufactured gases of recent origin. These include chlorofluorocarbons (CFCs) and **halons**. These gases have been extremely useful in many ways – for example, as refrigerants, for extinguishing fires, for making foams and plastics, and for use in aerosol spray cans. This is because they have some valuable properties: they are stable, non-flammable and non-toxic. Unfortunately, their stability means that they can persist a long time in the atmosphere and can thus reach the ozone layer without being destroyed. Once they are in the ozone layer, UV radiation from the sun starts to break them down. This sets off a chain of chemical reactions in which reactive chlorine atoms are released. These act as a **catalyst** causing ozone (O_3) to be converted into oxygen (O_2) (figure III.11).

Global production of CFC gases increased greatly during the 1960s, 1970s and 1980s, from around 180 million kg per year in 1960 to nearly 1,100 million kg per year in 1990. However, in response to the thinning of the ozone layer, many governments signed an international agreement called the Montreal Protocol in 1987. This pledged them to a rapid phasing out of CFCs and halons. Production has since dropped substantially. However, because of their stability, these gases will persist in the atmosphere for decades or even centuries to come. Even with the most stringent controls that are now being considered, it will be the middle of the twenty-first century before the chlorine content of the stratosphere falls below the level that triggered the formation of the Antarctic 'ozone hole' (see below) in the first place.

Some thinning of the ozone layer may result from time to time from natural rather than anthropogenic processes. A possible factor may be the pollution of the stratosphere with particulate material (aerosols) emitted by volcanic eruptions such as that of Mt Pinatubo in June 1991.

The most drastic decline in stratospheric ozone has been over Antarctica. This has led to the formation of the 'ozone hole' which expanded to an area

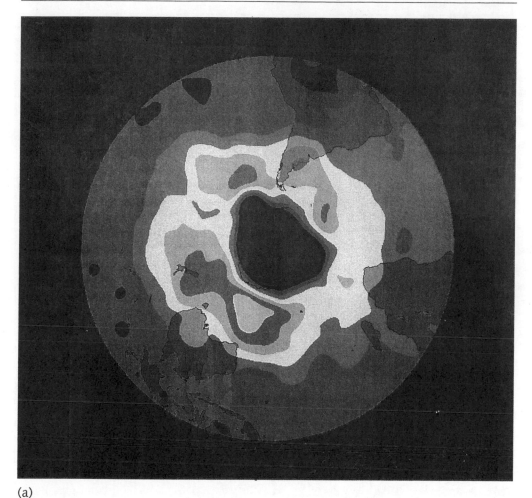

(a)

Plate III.5   (a) The Antarctic 'ozone hole' from space, 8 October 1995; *overleaf* (b) the Northern Hemisphere 'ozone hole', 12 March 1995. The colours represent ozone concentrations in Dobson units. (NOAA/Science Photo Library)

Table III.4   Trends in stratospheric ozone, 1979–1991 (% per decade)

	December–March	May–August	September–November
Satellite-derived data			
45°N	−5.6 ± 3.5	−2.9 ± 2.1	−1.7 ± 1.9
Equator	+0.3 ± 4.5	+0.1 ± 5.2	+0.3 ± 5.0
45°S	−5.2 ± 1.5	−6.2 ± 3.0	−4.4 ± 3.2
Land-based data			
26°N–64°N	−4.7 ± 0.9	−3.3 ± 1.2	−1.2 ± 1.6

Source: Tolba and El-Kholy (eds) (1992), table 2, p. 50.

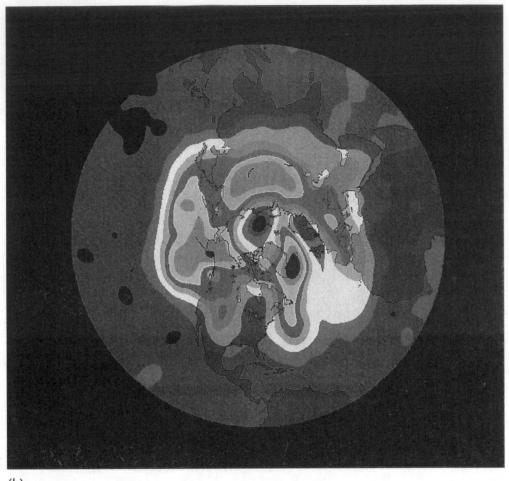

(b)

Figure III.11, *opposite* (a) The naturally occurring chemical processes leading to the formation and decomposition of ozone in the atmosphere in the presence of ultraviolet radiation. (b) The decomposition of ozone initiated by chlorine atoms released during the breakdown of a commonly occurring, anthropogenically generated CFC believed to be harmful to the atmosphere ($CFCl_3$). Not all the two-atom (diatomic) molecules of oxygen combine to form ozone, and the free chlorine atoms that are liberated are potentially capable of initiating further reactions that lead to the breakdown of ozone. (c) Schematic diagram to show the principal sources of atmospheric ozone, and the main reactions that cause ozone depletion in the stratosphere.
Source: Pickering and Owen (1994).

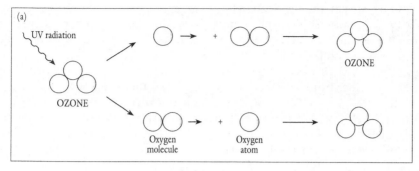

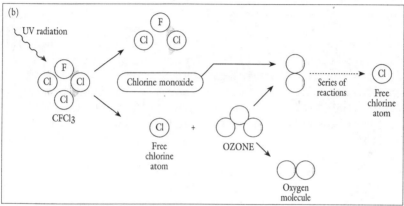

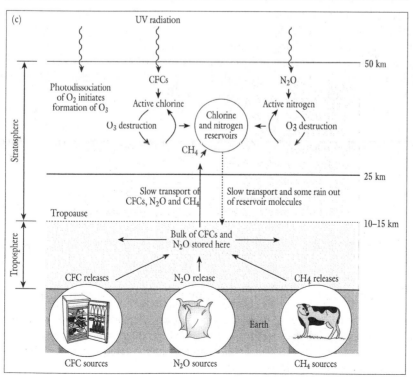

of 24 million sq km during September–October 1992 and again in the same months of 1993. Record low ozone levels of less than 100 ozone units were registered during a few days in October 1993. These compare with values from years before the ozone hole (1957–78) of 330–350 units.

The destruction of ozone is greatest over the Antarctic because of the unique weather conditions during the long, dark winter of the south polar regions. Strong winds circulate in a great vortex above the Antarctic, essentially isolating the polar stratosphere from the rest of the atmosphere. Under the very cold conditions, with temperatures below −80°C, ice clouds form, called polar stratospheric clouds. These provide ideal conditions for the transformation of chlorine (derived from the breakdown of CFCs and halons) into potentially reactive compounds. When sunlight returns in the spring months, UV radiation from the sun triggers the reaction between these chlorine compounds and ozone, thereby leading to ozone destruction.

No such clear ozone hole develops over the Arctic because the more complex arrangement of land and sea here leads to a less well developed vortex system of winds. In addition, the winter stratosphere at the North Pole tends to be warmer than its southern counterpart. This means that polar stratospheric clouds are usually less abundant. Nonetheless, ozone depletion does seem to have occurred, producing an ozone 'crater' rather than a hole.

Paradoxically, while ozone levels may be dropping in the stratosphere, at lower levels in the atmosphere they are increasing. This tropospheric ozone is produced by the action of sunlight on the nitrogen oxides and hydrocarbons that are emitted in fossil fuel exhaust gases. Such **photochemical reactions**, as they are called, are particularly serious in some great cities, like Los Angeles, where the high density of vehicles, the frequent occurrence of sunshine and the favourable topography lead to high concentrations of a soup of photooxidant gases. Research in both America and Europe has established that extensive formation of tropospheric ozone also frequently occurs in Northern Hemisphere mid-latitudes in the summer in non-urban areas, most noticeably downwind of cities and major industrial regions. The problem is regional rather than merely urban.

High levels of ozone concentration have several serious consequences. Humans suffer from eye irritation, respiratory complaints and headaches. Ozone is also potentially toxic to many species of coniferous trees, herbaceous plants and crops at concentrations not far above the natural background level. Rigorous controls on vehicle emissions can greatly reduce the problem. Such measures are now being implemented in California. Indeed, as figure III.12 shows, in spite of a hefty increase in both population and the number of motor vehicles in the Los Angeles area since 1970, peak ozone levels have declined very markedly and the area subjected to high ozone concentrations has shrunk (Lents and Kelly, 1993).

FURTHER READING

Gribbin, J., 1988, *The Hole in the Sky: Man's Threat to the Ozone Layer*. London: Corgi Books.
An introductory treatment for the general public by a well-known scientific journalist.

Mintzer, I. M. and Miller, A. S., 1992, Stratospheric ozone depletion: can we save the sky? In *Green Globe Yearbook 1992*, 83–91. Oxford: Oxford University Press.
A more recent general discussion of the causes and consequences of the ozone hole and what can be done to deal with it.

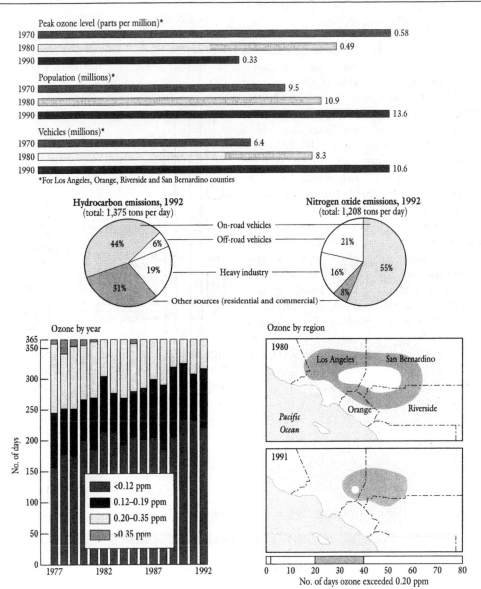

Figure III.12 Air pollution in the Los Angeles area, 1970s–1990s
Source: After Lents and Kelly (1993), p. 22.

8   ACID DEPOSITION

Rain is slightly acid under natural conditions because it contains some dissolved gases, including carbon dioxide (CO_2), sulphur dioxide (SO_2) and nitrogen oxides (NOx). These gases are naturally pre-sent in the air. Under natural conditions rain has a pH of around 5.65 (figure III.13). The term 'acid rain' was introduced as long ago as the 1850s for rain which has a pH of less than 5.65. Such rain has become more than usually acid because of air pollution. Two of the chemical reactions involved are shown in figure III.14.

Some scientists prefer the term 'acid deposition' to 'acid rain', for not all environmental acidification is caused by acid rain in the narrow sense. Acidity can reach the ground surface without the assistance of water droplets as particulate matter or gases. This is termed 'dry deposition'. Furthermore, there are various different types of wet deposition: mist, fog, hail, sleet and snow, as well as rain itself.

As a result of air pollution, precipitation in many parts of the world has pH values far below 5.65. Snow and rain in the north-east USA have been known to have pH values as low as 2.1. In the eastern

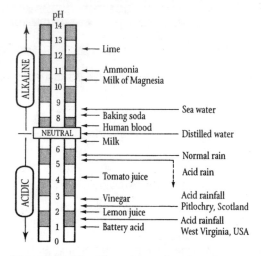

Figure III.13   The pH scale, showing the pH level of acid rain in comparison with that of other common substances
Source: Kemp (1994), figure 4.1.

(1) Sulphurous and sulphuric acids

SO_2 is emitted from natural and anthropogenic sources and dissolves in cloud water to produce sulphurous acid:

$$SO_2 + H_2O \longrightarrow H_2SO_3 \longrightarrow H^+ + HSO_3^-$$

Sulphurous acid can be oxidized in the gas or aqueous phase by various oxidants:

$$SO_2 \xrightarrow{\text{oxidant}} SO_3$$

Aqueous sulphur trioxide forms sulphuric acid:

$$SO_3 + H_2O \longrightarrow H_2SO_4 \longrightarrow H^+ + HSO_4^- \longrightarrow 2H^+ + SO_4^{2-}$$

(2) Nitrous and nitric acids

NO and NO_2 (collectively known as NOx) are produced by combustion processes and lightning. Nitric and nitrous acids may be produced:

$$2NO_2 + H_2O \longrightarrow HNO_3 + HNO_2$$

Figure III.14   Chemical reactions producing acid deposition
Source: Modified after Mannion (1992), fig. 11.2.

Table III.5 Main sources of acid gases in the UK in 1990		
	Annual emissions (000 tonnes)	% of UK total
Nitrogen oxides		
Road transport	1,400	51
Power stations	780	28
Industry	270	9
Sulphur dioxide		
Power stations	2,700	72
Industry	710	19

Source: Department of Environment figures.

USA as a whole the average annual acidity values of precipitation tend to be around pH4. The pH scale is logarithmic, so a decrease of one pH unit represents a tenfold increase in acidity. Thus pH4 is ten times more acidic than pH5. The main gases responsible for this state of affairs are the sulphur oxides and nitrogen oxides emitted from fossil fuel combustion (see table III.5). As a general rule, sulphur oxides have the greatest effect, and are responsible for about two-thirds of the problem. However, in some regions, such as Japan and the west coast of the USA, the nitric acid contribution may well be of relatively greater importance.

Whichever of these gases is most important, most acidification has occurred in the industrialized lands of the Northern Hemisphere. It is here that emissions of sulphur and nitrogen oxides are highest, because of high rates of fossil fuel combustion by a range of sources, notably industries, cars and power stations. However, the pollutants that cause acidification can be transported over long distances by the wind. The acidification of Scandinavia, for example, has been attributed in part to emissions from Britain. Similarly, Canada receives much of its acid deposition from the industrial heartland and the Ohio River

Valley region of the USA. Recent estimates of global emissions of sulphur suggest that anthropogenic sources now account for 55–80 per cent of the combined total (anthropogenic and natural), and that over 90 per cent of emissions from anthropogenic sources originate in the Northern Hemisphere.

The effects of acid deposition are greatest in those areas which have high levels of precipitation (causing more acidity to be transferred to the ground) and those which have base poor (acidic) rocks which cannot neutralize the deposited acidity.

Some of the most persuasive evidence for long-term increases in acid deposition is provided by what is called the **palaeolimnological** approach. In this approach, past environmental information is obtained by looking at the changes in the faunal and floral content of cores of sediment taken from the floors of lakes. The record provided by **diatoms** is especially useful, for these algae are excellent indicators of water chemistry. The composition of fossil assemblages retrieved from dated cores can be used to reconstruct changes in water pH. In Britain, at sensitive sites, pH values of lake waters were close to 6.0 before 1850, but since then pH declines

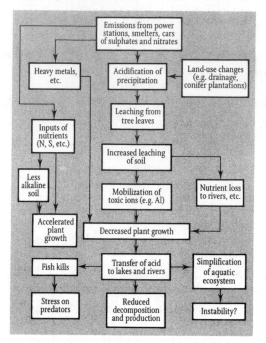

Figure III.15 Pathways and effects of acid precipitation through different components of the ecosystem, showing some of the adverse and beneficial consequences

have varied between 0.5 and 1.5 units overall.

Acid precipitation has many ecological consequences (figure III.15). One harmful effect is a change in soil character. The high concentration of hydrogen ions in acid rain causes accelerated leaching of essential nutrients, making them less available for plant use. Furthermore, aluminium and some heavy metal ions become more soluble at low pH values and may have toxic effects on plants and aquatic organisms. Forest growth can also be affected. Acid rain can damage foliage, increase susceptibility to disease, affect germination and reduce nutrient availability (figure III.16).

Particular fears have been expressed about the possible effects of acid deposition on aquatic ecosystems, especially on fish populations. Many fish are intolerant of low pH values (table III.6). Fishless lakes are now common in areas like the Adirondacks in the north-east USA. Fish may also be adversely affected by the increasing amounts of toxic metal ions (e.g. aluminium) in surface waters.

Changes in land use can also make surface waters more acid. Modern forestry practices, for example, contribute to the problem, with drainage, clear felling and then the planting of monocultures of fast-growing species such as conifers. In these conditions acidic leaf litter builds up more speedily than might be the case naturally. This can add to the nutrient leaching effects of acid rain. Tall trees are also more effective at 'scavenging' airborne pollutants from clouds than, say, upland grassland. This serves to increase the amount of pollution deposited.

Another adverse effect of acid rain is the weathering of buildings, particularly those made from limestone, marble and sandstone. For example, sulphate-rich precipitation reacts with limestone to bring about chemical changes (e.g. the formation of calcium sulphate, or **gypsum**) which cause blistering, while the low pH values encourage the dissolution of the limestone. Many of the great cathedrals of Europe have been attacked in this way.

Various methods are used to try to reduce the damaging effects of acid deposition. One of these is to add powdered limestone to lakes to increase their pH values. However, the only really effective and practical long-term treatment is to curb the emissions of the offending gases. This can be achieved in a variety of ways: by reducing the amount of fossil fuel combustion; by using less sulphur-rich fossil fuels; by using alternative energy sources that do not produce nitrate or sulphate gases (e.g. hydropower or nuclear power); and by removing the pollutants before they reach the atmosphere. For example, after combustion at a power station, sulphur can be

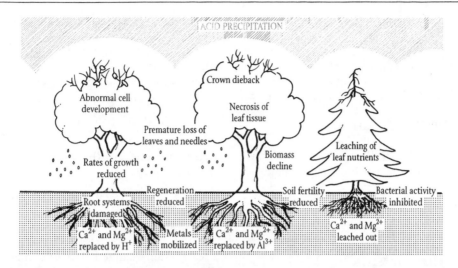

Figure III.16   The impact of acid precipitation on the terrestrial environment
Source: Various sources in Kemp (1994), figure 4.11.

Table III.6	Ecological effects of water pH on European freshwater fish
pH range	*Effects*
3.0–3.5	Unlikely that any fish can survive for more than a few hours.
3.5–4.0	This range lethal to salmonids. Tench, roach, pike and perch can survive.
4.0–4.5	Likely to be harmful to salmonids, tench, bream, roach, goldfish and common carp. Fish can become acclimatized to these levels.
4.5–5.0	Likely to be harmful to adults, eggs and fry of salmonids. Can harm common carp.
5.0–6.0	Unlikely to harm fish, unless free carbon dioxide concentration greater than 20 mg/l or water contains iron salts.
6.0–6.5	Unlikely to harm fish unless free carbon dioxide in excess of 100 mg/l.
6.5–9.0	Harmless to fish.
9.0–9.5	Likely to harm salmonids and perch if present for a long time.
9.5–10.0	Lethal to salmonids over prolonged periods.
10.0–10.5	Can be withstood for only short periods by roach and salmonids.
10.5–11.0	Rapidly lethal to salmonids. Prolonged exposure lethal to carp, tench, goldfish and pike.
11.0–11.5	Rapidly lethal to all species of fish.

Source: Gleick (1993), table F2.

removed ('scrubbed') from flue gases by a process known as flue gas desulphurization (FGD), in which a mixture of limestone and water is sprayed into the flue gas which converts the sulphur dioxide (SO_2) into gypsum (calcium sulphate). NOx in flue gas can be reduced by adding ammonia and passing it over a catalyst to produce nitrogen and water (a process called selective catalytic reduction or SCR). NOx produced by cars can be reduced by fitting a catalytic converter.

FURTHER READING

Park, C. C., 1987, *Acid Rain: Rhetoric and Reality*. London: Methuen.
A general introduction that provides a useful overview.

Wellburn, A., 1988, *Air Pollution and Acid Rain: The Biological Impact*. London: Longman.
A more advanced treatment with a strong biological emphasis.

9   CONCLUSION

Changes in the composition of the Earth's atmosphere as a result of human emissions of trace gases, and changes in the nature of land cover, have caused great concern in recent years. Global warming, ozone depletion and acid rain have become central issues in the study of environmental change. Although most attention is often paid to climatic change resulting from greenhouse gases, there is a whole series of other mechanisms which have the potential to cause climatic change. Most notably, we have pointed to the importance of other changes in atmospheric composition and properties, whether these are caused by aerosol generation or albedo change.

However, the greenhouse effect and global warming may prove to have great significance for the environment and for human activities. Huge uncertainties remain about the speed, degree, direction and spatial patterning of potential change. Nonetheless, if the Earth warms up by a couple of degrees over the next hundred or so years, the impacts, some negative and some positive, are unlikely to be trivial. The box in this part on the effects of warming on agriculture in the UK indicates this clearly.

For many people, especially in cities, the immediate climatic environment has already been changed. Urban climates are different in many ways from those of their rural surroundings. The quality of the air in many cities has been transformed by a range of pollutants, and we have pointed to the particularly serious levels of pollution that developed in Eastern Europe and in South Africa. Conversely, we have pointed out that under certain circumstances clean air legislation and other measures can cause rapid and often remarkable improvements in this area.

The same is true of two major pollution issues – ozone depletion and acid deposition. Both processes have serious environmental consequences and their effects may remain with us for many years, but both can be slowed down or even reversed by regulating the production and output of the offending gases.

The human impacts on the atmosphere discussed in this part of the book show clearly how different the impacts can be on different parts of the population, and also how impacts can spread widely, often affecting people a long way from the source of the problem. Furthermore, impacts on the atmosphere show forcefully the interlinked nature of environmental systems, and the knock-on effects of many atmospheric changes on the biosphere, fresh waters and land surface.

KEY TERMS AND CONCEPTS

acid rain
aerosols
albedo
dust bowl
global warming
greenhouse effect

land cover
ozone hole
stratospheric ozone
thermal pollution
tropospheric ozone
urban heat island

POINTS FOR REVIEW

What forces could (a) cause future climate to cool and (b) cause future climate to become warmer?

Can humans change regional and global precipitation patterns?

Is global warming an important environmental issue, and, if so, why?

Ozone concentrations are increasing in many cities but decreasing in the stratosphere. Why should this be?

Is acid rain an increasingly important or decreasingly important environmental issue? Defend your answer.

PART IV

The Waters

1 INTRODUCTION

In a recent review of the world's fresh-water resources, Gleick (1993, p.1) summed up the importance of water in a few clear sentences:

Fresh water is a fundamental resource, integral to all environmental and societal responses. Water is a critical component of ecological cycles. Aquatic ecosystems harbour diverse species and offer many valuable services. Human beings require water to run industries, to provide energy, and to grow food.

Because water is so important to human affairs, humans have sought to control water resources in a whole variety of ways. Also, because water is such an important part of so many natural and human systems, its quantity and quality have undergone major changes as a consequence of human activities. Again we can quote Gleick (1993, p. 3):

As we approach the 21st century we must now acknowledge that many of our efforts to harness water have been inadequate or misdirected ... Rivers, lakes, and groundwater aquifers are increasingly contaminated with biological and chemical wastes. Vast numbers of people lack clean drinking water and rudimentary sanitation services. Millions of people die every year from water-related diseases such as malaria, typhoid, and cholera. Massive water developments have destroyed many of the world's most productive wetlands and other aquatic habitats.

In this chapter we look at some of the ways in which the quantity and quality of water have been modified in some of the world's freshwater systems – rivers, groundwater and lakes. Table IV.1 summarizes some of the hydrological effects of land-use change and dem[...] great number and diversity.

2 RIVER REGULATION

In recent decades human demand for fresh water has increased rapidly. Global water use has more than tripled since 1950, and now stands at 4,340 cu km per year – equivalent to eight times the annual flow of the Mississippi River. Annual irretrievable water losses have increased about sevenfold this century.

One major way of regulating rivers is to build dams. Many new large dams have been built in the twentieth century, especially between 1945 and the early 1970s, and there are now more than 36,000 dams around the world. As table IV.2 shows, large dams (i.e. more than 15 metres high) are still being constructed in substantial numbers, especially in Asia. In the late 1980s some 45 very large dams (more than 150 metres high) were being built. Indeed, one of the most striking features of newly constructed dams and reservoirs is that they have become increasingly large (table IV.3).

Most dams achieve their aim, which is to regulate river **discharge**. They are also highly successful in meeting the needs of surrounding communities: millions of people depend upon them for survival, welfare and employment. However, dams have many environmental consequences that may or may not have been anticipated (figure IV.1). Some of these are dealt with in greater detail elsewhere (e.g. salinity, in part V, section 5).

The River Nile, before and after the construction of the great Aswan High Dam in Egypt (table IV.4), provides a good example of how dams retain sediment. Until the dam was built, concentrations of silt were high in the late summer and autumn period of high flow on the Nile. Since the dam has been finished, the silt

Table IV.1 Summary of the major hydrological effects of land-use changes

Land use change	Hydrological component affected	Principal hydrological processes involved
Afforestation (deforestation has the opposite effects in general)	Annual flow	Increased interception in wet periods Increased transpiration in dry periods
	Seasonal flow	Increased interception and increased dry period transpiration reduce dry season flow Drainage improvements associated with planting may increase dry season flows Cloud water (mist and fog) deposition on trees will augment dry season flows
	Floods	Interception reduces floods by removing a portion of the storm rainfall, and allowing soil moisture storage to increase Management activities, such as drainage, construction, all increase floods
	Water quality	Leaching of nutrients reduced, as surface runoff reduced and less application of fertilizer Deposition of atmospheric pollutants increased because of larger exposed surface area of trees
	Erosion	High infiltration rates in natural, mixed forests reduce surface runoff and erosion Slope stability enhanced by reduced soil pore water pressure and binding effect of tree roots Wind throw of trees reduces slope stability Management activities (construction, drainage) all increase erosion
	Climate	Increased evaporation and reduced sensible heat fluxes from forests affect climate
Agricultural intensification	Water quantity	Alteration of transpiration rates affects runoff Timing of storm runoff altered through land drainage
	Water quality	Application of inorganic fertilizers adds nutrients Pesticide application poses health risks to humans and animals Farm wastes pollute surface and groundwater where inadequate disposal of organic and inorganic wastes

Table continues opposite

Table IV.1 Continued

Land use change	Hydrological component affected	Principal hydrological processes involved
	Erosion	Cultivation without proper soil conservation measures, and uncontrolled grazing, increase erosion
Draining wetlands	Seasonal flow	Lowering of water table may induce soil moisture stress, reduce transpiration and increase dry season flows Initial dewatering on drainage will increase dry season flows
	Annual flow	Initial dewatering on drainage will increase annual flow Afforestation after drainage will reduce annual flow
	Floods	Drainage method, soil type and channel improvement will all affect flood response
	Water quality	**Redox potentials** altered, leading to peat decomposition, acidification and increased organic loads in runoff New drainage systems intercepting mineral horizons will reduce acidity
	Carbon balance	Accumulating peat bogs are sink for atmospheric CO_2

Source: Adapted from Calder (1992), table 13.1.1.

Table IV.2 Number of large dams (over 15 metres high) 1950 and 1986

Continent	1950	1986	Under construction 31 Dec. 1986
Africa	133	885	58
Asia	1,562	23,555	615
of which in China	8	18,820	183
Australasia/Oceania	151	497	25
Europe	1,323	4,077	230
North and Central America	2,099	6,663	39
South America		885	69
TOTAL	5,268	36,562	1,036

Source: Data provided by UNEP.

Table IV.3   World's 20 largest reservoirs, by reservoir volume

Name[a]	Country	Capacity (million cu metres)	Year completed
Owen Falls[b]	Uganda	204,800	1954
Bratsk	FSU[c]	169,000	1964
High Aswan	Egypt	162,000	1970
Kariba	Zimbabwe–Zambia	160,368	1959
Akosombo	Ghana	147,960	1965
Daniel Johnson	Canada	141,851	1968
Guri	Venezuela	135,000	1986
Krasnoyarsk	FSU	73,300	1967
W. A. C. Bennett	Canada	70,309	1967
Zeya	FSU	68,400	1978
Cahora Bassa	Mozambique	63,000	1974
La Grande 2 Barrage	Canada	61,715	1978
La Grande 3 Barrage	Canada	60,020	1981
Ust-Ilim	FSU	59,300	1977
Boguchany	FSU	58,200	under construction
Kuibyshev	FSU	58,000	1955
Serra da Mesa	Brazil	54,400	under construction
Caniapiscau Barrage K A 3	Canada	53,790	1980
Bukhatarma	FSU	49,800	1960
Ataturk	Turkey	48,700	1990

[a] All these reservoirs have been constructed since the Second World War.
[b] Owen Falls' capacity is not fully related to construction of a dam; the major part of it is a natural lake.
[c] Former Soviet Union.
Source: Modified from Gleick (1993), table G9.

load is lower throughout the year and the seasonal peak is removed. The Nile now only transports about 8 per cent of its natural sediment load below the Aswan High Dam. This figure is exceptionally low, probably because of the great length and size of Lake Nasser, the reservoir behind the dam. Other rivers for which data are available carry between 8 per cent and 50 per cent of their natural suspended loads below dams.

The removal of sediment from the Nile has various possible consequences. These include a reduction in flood-deposited nutrients on fields; less nutrients for fish in the south-east Mediterranean Sea; accelerated erosion of the Nile Delta; and accelerated riverbed erosion, since less sediment is available to cause bed **aggradation**. The last process is often called 'clear-water erosion'. It may speed up the rate at which streams cut back into their banks in an upstream direction. It may also cause groundwater tables to become lower and undermine bridge piers and other structures downstream of the dam. On the

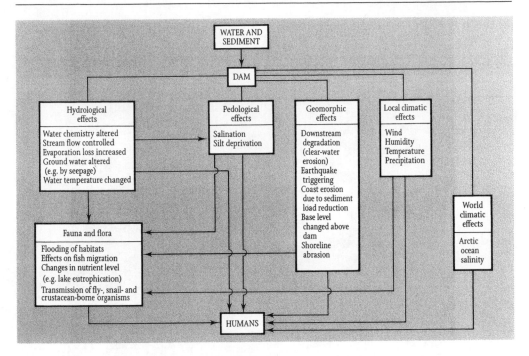

Figure IV.1 Generalized representation of the possible effects of dam construction on human life and various components of the environment

Table IV.4 Silt concentrations in the Nile at Gaafra before and after the construction of the Aswan High Dam (ppm)

Jan.	Feb.	March	April	May	June	July	Aug.	Sep.	Oct.	Nov.	Dec.
Before (averages for the period 1958–63)											
64	50	45	42	43	85	674	270	242	925	124	77
After											
44	47	45	50	51	49	48	45	41	43	48	47
Ratio of before to after											
1.5	1.1	1.0	0.8	0.8	1.7	14.0	60.0	59.1	21.5	2.58	1.63

Source: Abu-Atta (1978), p. 199.

other hand, in regions such as northern China, where modern dams trap silt, the cutting-out of the river channel downstream may alleviate the strain on **levées** and so lessen the expense of strengthening or heightening the levées.

However, clear-water erosion does not always follow from dam construction. In some rivers, before a dam was built, the sediment brought into the main stream by steep tributaries was carried away by floods. Once the dam is built, these floods no longer happen and so the sediment accumulates as large fans of sand or gravel below each tributary mouth. The bed of the main stream is raised and any water

Plate IV.1   The Sagan River in southern Ethiopia. The dark brown colour of this river is caused by its large load of sediment derived from accelerated erosion upstream. This renders the river much less suitable as a source of drinking water. (A. S. Goudie)

intakes, towns or other structures that lie alongside the river may be threatened by flooding or channel shifting across the accumulating wedge of sediment.

Some landscapes are almost dominated by dams, canals and reservoirs. Probably the most striking example of this is the 'tank' landscape of south-east India where myriads of little streams and areas of overland flow have been dammed by small earth structures to produce what Spate (Spate and Learmonth, 1967, p. 778) has likened to 'a surface of vast overlapping fish-scales'.

In the northern part of the Indian subcontinent, in Sind, the landscape changes brought about by hydrology are no less striking. Here, the mighty snow-fed Indus River is controlled by large embankments (*bunds*) and interrupted by great dams. Its waters are distributed over thousands of square kilometres by a network of canals that has evolved over the past 4,000 years.

Another direct means of river manipulation is channelization. This involves constructing embankments, dikes, levées and floodwalls to confine floodwaters; and improving the ability of channels to transmit floods by enlarging their capacity through straightening, widening, deepening or smoothing.

Some of the great rivers of the world are now lined by extensive embankment systems such as those that run for more than 1,000 km alongside the Nile, 700 km along the Hwang Ho in China, 1,400 km by the Red River in Vietnam, and over 4,500 km in the Mississippi Valley. Like dams, embankments and other such structures often fulfil their purpose but may also create environmental problems and have some disadvantages. For example, they reduce natural storage for floodwaters, both by preventing water from spilling on to much of the floodplain, and, where impermeable floodwalls are used, by not

allowing water to be stored in the banks. The flow of water in tributaries may also be constrained. Occasionally, embankments may exacerbate the flood problem they were designed to reduce. This can happen where the barriers downstream of a breach prevent floodwater from draining back into the channel once the peak has passed.

Channel improvement, designed to improve water flow, may also have unforeseen or undesirable effects. For example, the more rapid movement of water along improved sections of a river channel can aggravate flood peaks further downstream and cause excessive erosion. The lowering of water tables in the 'improved' reach may cause overdrainage of adjacent agricultural land. In such cases, sluices need to be constructed in the channel to maintain levels. On the other hand, channels lined with impermeable material may obstruct soil water movement (interflow) and shallow groundwater flow, thereby causing surface saturation.

Channelization may also have various effects on fauna. These may result from faster water flow, reduced shelter in the channel bed, and reduced food supplies due to the destruction of overhanging bank vegetation. If channelization of rivers were carried out in large swamps, like those of the Sudd in Sudan or the Okavango in Botswana, where plans to do so exist, it could completely transform the whole character of the swamp environment.

Another type of channel modification is the construction of bypass and diversion channels, either to carry excess floodwater or to enable irrigation to take place. The use of such channels may be as old as irrigation itself. They may contribute to the salinity problems encountered in many irrigated areas (see part V, section 5).

Deliberate modification of a river regime can also be achieved by long-distance inter-**basin** water transfers (Shiklomanov, 1985; and see section 9 below). Such transfers are necessitated by the unequal spatial distribution of water resources and by the increasing rates of water consumption. At present, the world water consumption for all needs is 4,340 cu km per year, nine times what it was at the beginning of the twentieth century. By the year 2000 it is expected to be 6,000 sq km per year. The total volume of water in the various transfer systems in operation and under construction throughout the world at present is about 300 sq km per year. The greatest volumes of transfers take place in Canada, the former USSR, the USA and India.

It is likely that many even greater schemes will be constructed in future decades. Route lengths of some hundreds of kilometres will be common, and the water balances of many rivers and lakes will be transformed. (See section 9 below for what has already happened to the Aral Sea.)

A human activity that affects many coastal portions of rivers, or estuaries, is dredging. The effects of dredging can be as complex as the effects of dams and reservoirs upstream (La Roe, 1977). Dredging may be performed to create and maintain canals, navigation channels, turn-ing basins, harbours and marinas; to lay pipelines; and to obtain a source of material for filling or construction. The ecological effects of dredging are various. In the first place, filling directly disrupts habitats like salt marshes. Second, the large quantities of suspended silt generated can physically smother plants and animals that live on river and estuary beds; smother fish by clogging their gills; reduce photosynthesis through the effects of **turbidity**; and lead to **eutrophication** by releasing large quantities of nutrients. Likewise, the destruction of marshes, mangroves and sea grasses by dredging and filling can result in the loss of these natural purifying systems (see part II, section 9 on wetlands). The removal of vegetation may also cause erosion. Moreover, as silt deposits stirred up by dredging accumulate elsewhere in the estuary they tend to create a 'false bottom'. The dredged bottom, with its shifting, unstable sediments, is recolonized by fauna and flora only slowly, if at all. Furthermore, dredging tends to change the configuration of currents and the rate of freshwater drainage, and may provide avenues for salt-water intrusion.

FURTHER READING

Brookes, A., 1985, River channelization: traditional engineering methods, physical consequences, and alternative practices. *Progress in Physical Geography* 9, 44–73.
An advanced review by a leading authority.

Gleick, P. H. (ed.), 1993, *Water in Crisis: A Guide to the World's Freshwater Resources*. New York: Oxford University Press.
An invaluable compendium of information on all aspects of water use and misuse. It contains many useful tables of data.

Gregory, K. J., 1985, The impact of river channelization. *Geographical Journal* 151, 53–74.
A useful overview in a relatively accessible journal.

Petts, G. E., 1985, *Impounded Rivers: Perspectives for Ecological Management*. Chichester: Wiley.
An advanced textbook that looks at the large range of consequences of dam construction.

Modification of the Colorado River, USA

Plate IV.2 The Hoover Dam on the Colorado River, Arizona, USA. The flow of the river, and its sediment load, are now almost totally controlled. (Trip/M. Lee)

The Colorado River in the American West (figure IV.2(a)), which flows through the Grand Canyon, has been at several points dammed to control floods, generate electricity and provide water for irrigation. Among the major dams are the Hoover and Glen Canyon dams, both over 200 m high (figure IV.2(b)). They have caused radical adjustments in the hydrological regime. Flood peaks are reduced as a flood control strategy, and water is released at times of low flow. Discharge varies rapidly in response to fluctuations in the need for hydropower during the course of a day. The high dams trap most of the sediment carried by the river, so that downstream discharges are largely sediment-free. In the Colorado River this combination of impacts has changed a natural river with very large spring floods, lower summer flows and little daily variation of sediment-laden waters into a highly controlled system with only modest flood peaks in spring, relatively high summer flows, and drastic daily variation of discharges of clear water. Indeed, at its seaward end the Colorado has been totally transformed. Prior to 1930, before the dams were built, it carried around 125–130 million tons of suspended sediment per year to its delta at the head of the Gulf of California (figure IV.2(c)). Now the Colorado discharges neither sediment nor water to the sea. Upstream,

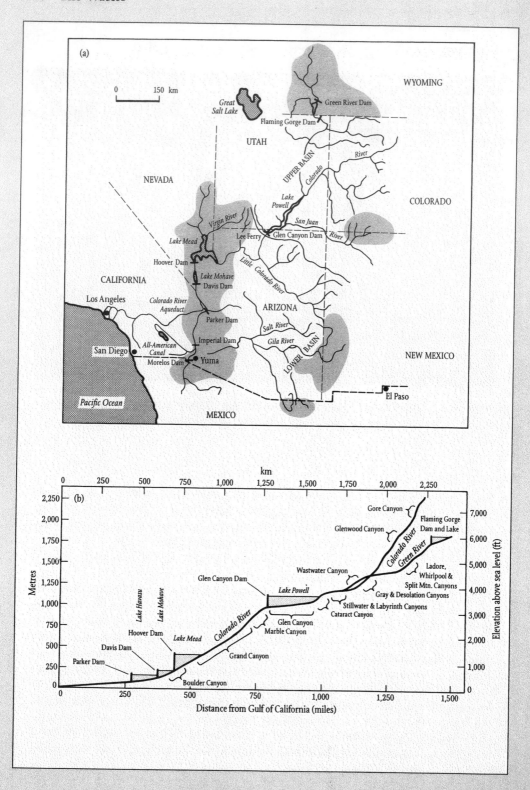

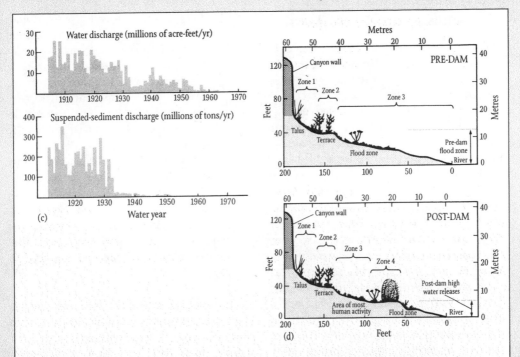

Figure IV.2   (a) The Colorado River basin. (b) Profiles of the Green and Colorado Rivers, showing locations of dams, reservoirs and whitewater canyons. (c) Historical sediment and water discharge of the Colorado River. (d) Pre- and post-dam riparian vegetation in the Grand Canyon downstream from Glen Canyon dam. Vegetation zones: 1. stable desert vegetation; 2. stable woody vegetation; 3. unstable zone; 4. new riparian vegetation, phreatophytes
Sources: (a) After Schwarz et al. (1990); (b) After various sources in Graf (1985) fig. 13; (c) After Schwarz et al. (1990); (d) After Graf (1985), from original by S. W. Carothers.

in the vicinity of the Grand Canyon, **riparian** vegetation communities have been completely changed since the construction of the Glen Canyon dam (figure IV.2(d)).

Further reading

Graf, W. L., 1985, *The Colorado River: Instability and Basin Management*. Washington DC: Association of American Geographers.

3 FORESTS AND RIVER FLOW

When George Perkins Marsh wrote his remarkable book *Man and Nature* in 1864, one of the main themes which concerned him was the consequences of forest removal. In the early twentieth century, scientists in America began to measure the effects of forest removal on stream discharges. To do this, they used what is sometimes called the 'paired watershed' technique. First, they compared the flows out of two similar watersheds (**catchments**) over a period of years. Then they clear-felled one of the watersheds to see how that basin responded in comparison with the unchanged, 'control' valley. The pioneering study, at Wagon Wheel Gap in Colorado, USA, in 1910, revealed that the clear-felled valley yielded 17 per cent more waterflow than would have been expected if it had remained unchanged, like the control valley. Subsequent studies in the tropics have indicated that clear-felling can lead to mean annual stream flow increases equivalent to about 400–450 mm of rainfall.

There are many reasons why the removal of a forest cover and its replacement with pasture, crops or bare ground have such important effects on stream flow. A mature forest probably intercepts a higher proportion of rainfall, tends to reduce rates of overland flow, and promotes soils with a higher **infiltration** capacity and better general structure. All these factors will tend to produce both a reduction in overall runoff levels and less extreme flood peaks, though this is not invariably the case.

Reforestation of abandoned farmlands reverses the effects of deforestation: increased interception of rainfall and higher levels of evapotranspiration can cause a decline in water yield to rivers. This can cause problems for human activities.

Reviews of catchment experiments from many parts of the world have pointed to two conclusions:

- Pine and eucalypt forest types cause an average change of 40 mm in annual flow for a 10 per cent change in cover with respect to grasslands; that is, a 10 per cent increase in forest cover on grassland will decrease annual flow by 40 mm, and a 10 per cent decrease in cover will increase annual flow by the same amount.
- The equivalent effect on annual flow of a 10 per cent change in cover of deciduous hardwood or scrub is 10–25 mm; that is, if 10 per cent of a grassland catchment is converted to hardwood trees or scrub vegetation, the annual runoff will decrease by 10–25 mm.

The increase in annual flow that results from tree or scrub removal tends to be most marked in two particular environments: those with very high rainfall and those with very low rainfall. In the former, evaporation from forest will tend to be higher than that from other land uses because of high levels of rainfall interception. In the latter, evaporation from forest is likely to be higher than that from other land uses because forests, composed of trees that have deep root systems, are better able to make use of soil and groundwater reserves.

Having discussed changes in annual flows, now let us turn to a consideration of how forest removal influences low season flows and flood peaks. The higher losses from forests in wet seasons from rainfall interception and increased losses in dry seasons from transpiration (because of trees' deeper root systems) both tend to increase soil moisture deficits in dry seasons compared to those under other land uses. On the other hand, in forests at high altitudes, where there is a lot of water deposition on to trees from clouds, this may provide a significant component of the dry season flows in rivers and also

Plate IV.3   A well-managed tea and rubber plantation in the Nilgiri hills, southern India. (A. S. Goudie)

increase runoff. The same applies in areas with high-intensity storms where high-intensity rainfall may lead to high levels of surface runoff. The higher infiltration rates under indigenous forest compared with other land uses may help soils and their below-ground aquifers to recharge themselves. In steeply sloped areas forests may have the additional benefit of reducing landslips (see part V, section 6) and preserving the soil aquifer which may be the source of dry season flows. Both these effects of afforestation may there-fore benefit stream flows in the low season.

When it comes to flood peaks there is still a great deal of controversy as to how important forest cover is with respect to the largest types of event. Some authors suggest that management practices associated with forestry (e.g. the building of roads, culverts and drainage ditches) or subsequent activities (e.g. grazing) which promote the flood, by causing compaction of the soil and reducing its infiltration capacity, increase this type of hazard.

FURTHER READING

I. R. Calder, 1992, Hydrologic effects of land-use change. In D. R. Maidment (ed.), *Handbook of Hydrology*, pp. 13.1–13.50. New York: McGraw Hill.
A lengthy and detailed summary of the available literature.

George Perkins Marsh: pioneer investigator of human impacts on forests and hydrology

George Perkins Marsh (1801–82) was born in Vermont, USA, and can be regarded as one of the most important pioneers of the conservation movement. In 1864 he wrote *Man and Nature*. This book was the product of two major influences on him: first, his upbringing in New England; and secondly, his experiences working for the US government in Turkey and elsewhere around the Mediterranean basin. In it he recognized how human occupation of the land had transformed it. This brief extract from *Man and Nature*, in which he deals with the consequences of forest destruction, gives a good indication of his clear and direct style:

With the disappearance of the forest, all is changed. At one season, the earth parts with its warmth by radiation to an open sky – receives, at another, an immoderate heat from the unobstructed rays of the sun. Hence the climate becomes excessive and the soil is alternately parched by the fervors of summer, and seared by the rigors of winter. Bleak winds sweep unresisted over its surface, drift away the snow that sheltered it from the frost, and dry up its scanty moisture. The precipitation becomes as regular as the temperature; the melting snows and vernal rains, no longer absorbed by a loose and bibulous vegetable mould, rush over the frozen surface, and pour down the valleys seaward, instead of filling a retentive bed of absorbent earth, and storing up a supply of moisture to feed perennial springs. The soil is bared of its covering of leaves, broken and loosened by the plough, deprived of the fibrous rootlets which hold it together, dried and pulverized by sun and wind, and at last exhausted by new combinations. The face of the earth is no longer a sponge, but a dust heap, and the floods which the waters of the sky pour over it hurry swiftly along its slopes, carrying in suspension vast quantities of earthly particles which increase the abrading power and mechanical force of the current, and, augmented by the sand and gravel of falling banks, fill the beds of the streams, divert them into new channels, and obstruct their outlets. The rivulets, wanting their former regularity of supply and deprived of the protecting shade of the woods, are heated, evaporated, and thus reduced in their summer currents, but swollen to raging torrents in autumn and spring. From these causes, there is a constant degradation of the uplands, and a consequent elevation of the beds of the watercourses and of lakes by the deposition of the mineral and vegetable matter carried down by the waters. The channels of great rivers become unnavigable, their estuaries are choked up, and harbors which once sheltered large navies are shoaled by dangerous sandbars. The earth, stripped of its vegetable glebe, grows less and less productive, and, consequently, less able to protect itself by weaving a new carpet of turf to shield it from wind and sun and scouring rain. Gradually it becomes altogether barren. The washing of the soil from the mountains leaves bare ridges of sterile rock, and the rich organic mould which covered them, now swept down into the dank low grounds, promotes a luxuriance of aquatic vegetation that breeds fever, and more insidious forms of mortal disease, by its decay, and thus the earth is rendered no longer fit for the habitation of man.

MARSH, 1861. PHOTOGRAPH BY BRADY

Courtesy of Frederick H. Meserve

Plate IV.4   George Perkins Marsh, author of *Man and Nature* (1864) and one of the major proponents of nature conservation.

Further reading

Marsh, G. P., 1864, *Man and Nature* (quoted from edition by D. Lowenthal, 1965, Cambridge, Mass.: Belknap Press of Harvard University Press, pp. 186–7).

4   THE HYDROLOGICAL RESPONSE TO URBANIZATION

The remarkable growth of the number and size of cities in recent decades has created many new impacts on water resources and distribution. For example, cities modify the precipitation characteristics of their immediate environs (see part III, section 5). They also can cause changes in water quality through thermal pollution (see section 8 below) and chemical pollution (see section 6 below). Moreover, the demand for water by city populations may be so great that groundwater is mined from city aquifers (see section 10 below) and large amounts are brought in by inter-basin water transfers. Los Angeles, for example, receives water from distant parts of northern California. In this section, however, we will concentrate on the effect of urbanization on river flow characteristics.

Research in various countries has shown that urbanization influences flood runoff. For example, figure IV.3 shows in a schematic way the hydrological changes resulting from urbanization in a part of Canada. These changes are caused mainly by the production of extended surfaces of tarmac, tiles and concrete. Because these impermeable surfaces have much lower infiltration capacities than rural, vegetated surfaces, they generate a rapid response to rainfall. This response is further accelerated by sewers, storm drains and the like which are very efficient at catching and transporting city rainfall. In general, the greater the area that is sewered, the greater is the discharge that will occur in any given period of time. In other words, the interval between flood events becomes progressively shorter. Moreover, peak discharges are higher, and occur sooner after runoff starts, in basins that have been affected by urbanization and sewer construction.

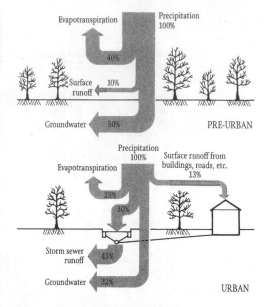

Figure IV.3   Hydrological changes in Ontario, Canada, caused by urbanization
Source: After OECD (1986), p. 43.

Table IV.5 shows the impact of different influences resulting from the urbanization process.

Some workers have found that urbanization has a proportionately greater effect on smaller flood events than on larger ones. In other words, the effects of urbanization appear less important as the size of the flood and the interval between floods increase. A probable explanation for this is that, during a severe and prolonged rainstorm, a rural catchment may become so saturated over large areas, and its channel network so extended, that it begins to behave almost as if it were an impervious urban catchment with a dense storm drain network. Under these conditions, a rural catchment produces floods rather similar to those of its urban counterpart. Also, in very large floods, subsurface drains in cities may not be large enough to take the volume of water, resulting in less rapid and lower discharge.

Table IV.5   Potential hydrological effects of urbanization

Urbanizing influence	Potential hydrological response
Removal of trees and vegetation	Decreased evapotranspiration and interception; increased stream sedimentation
Initial construction of houses, streets and culverts	Decreased infiltration and lowered groundwater table; increased storm flows and decreased base flows during dry periods
Complete development of residential, commercial and industrial areas	Decreased porosity, reducing time of runoff concentration, thereby increasing peak discharges and compressing the time distribution of the flow; greatly increased volume of runoff and flood damage potential
Construction of storm drains and channel improvements	Local relief from flooding; concentration of floodwaters may aggravate flood problems downstream

Source: Kibler (1982).

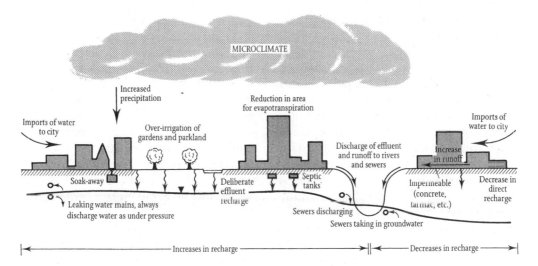

Figure IV.4   Urban effects on groundwater recharge
Source: After Lerner (1990), fig. 2.

Different cities, different construction methods, and other variable factors will all affect the response to rainfall inputs, and we should avoid overgeneralization. Urban groundwater provides an example. We have suggested that surface runoff is increased by the presence of impermeable surfaces. One consequence of this would be that less water went to recharge groundwater. However, there is an alternative point of view, namely that groundwater recharge can be accelerated in urban areas because of leaking water mains, sewers, septic tanks and soakaways (figure IV.4).

In cities in arid areas there is often no adequate provision for storm runoff, and the (rare) increased runoff from impermeable surfaces will infiltrate into the permeable surroundings. In some cities recharge may result from over-irrigation of parks and gardens. Indeed, where the climate is dry, or where large supplies of water are imported, or where pipes and drains are poorly maintained, groundwater recharge in urban areas is likely to exceed that in rural areas.

Further Reading

Lerner, D., 1990, *Groundwater Recharge in Urban Areas*, 59–65. IAHS Publication no. 198.
A cogent account of the role of groundwater in the urban environment.

5   Land Drainage

The drainage of wet soils has been one of the most successful ways in which rural communities have striven to increase agricultural productivity. It was, for example, practised centuries ago by the Etruscans, Greeks and Romans.

Large areas of marshland, floodplain and other wetlands have been drained to human advantage. When water is led away, the **water table** is lowered and stabilized, providing greater soil depth for plant rooting. Moreover, well-drained soils warm up earlier in the spring and thus permit crops to be planted, and to germinate, earlier. Farming is easier if the soil is not too wet, since the damage to crops by winter freezing may be reduced, undesirable salts are carried away from irrigated areas, and the general physical condition of the soil is improved. In addition, drained land tends to be flat and so is less prone to erosion and more amenable to mechanical cultivation. It will also be less prone to drought risk than certain other types of land. By reducing the area of saturated ground, drainage can alleviate flood risk in some situations by limiting the extent of a drainage basin that generates **saturation excess overland flow**, but this is an issue we shall return to later.

The most spectacular feats of drainage are the **arterial drainage** systems, involving the construction of veritable rivers and networks of large dikes, seen for example in the Netherlands and the Fenlands of eastern England. These have received much attention. However, more widespread than arterial drainage, and sometimes independent of it, is the drainage of individual fields. This is done either by surface ditching or by underdrainage with tile pipes and the like. In Finland, Denmark, Great Britain, the Netherlands, Hungary and the fertile Midwest of the USA, the majority of agricultural land is drained.

In Britain underdrainage was promoted by government grants and in the 1970s in England and Wales reached a peak of about 1 million hectares per year. More recently, government subsidies have been cut and the uncertain economic future of farming has led to a reduction in farm expenditure. Both tendencies have led to a reduction in the growth of field drainage, which is now being extended by only about 40,000 hectares per year (Robinson, 1990).

Drainage is a widespread practice which has many advantages and benefits. However, it can also have environmental costs. The first of these is related to a reduction in the extent of highly important wetland wildlife habitats (see part II, section 9). Marshes, fens and swamps are of major ecological significance for a wide range of species.

Secondly, the drainage of organically rich

Plate IV.5 Drainage maintenance on agricultural land in the Fenlands of eastern England at Spalding in Lincolnshire. (E.P.L./Richard Teeuw)

soils (such as those that contain much peat) can lead to the degradation and eventual disappearance of peaty materials, which in the early stages of post-drainage cultivation may be highly productive for agriculture. The lowering of the water table makes peats susceptible to **oxidation** and deflation (removal by wind) so that their volume decreases. One of the longest records of this process, and one of the clearest demonstrations of its efficacy, has been provided by the measurements at Holme Fen Post in the English Fenlands. Approximately 3.8 metres of subsidence occurred between 1848 and 1957, with the fastest rate occurring soon after drainage had been initiated (figure IV.5). The present rate averages about 1.4 cm per year. At its maximum natural extent, before the Middle Ages, the peat of the English Fenland covered around 1,750 sq km. Now only about one-quarter (430 sq km) remains.

Similar subsidence has taken place following drainage of portions of the Florida

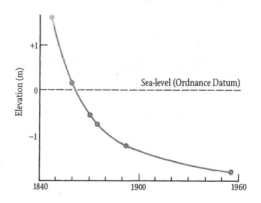

Figure IV.5 The subsidence of the English Fenlands peat in Holme Fen Post from 1842 to 1960 following drainage
Source: After Goudie (1993), fig. 6.8, from data in Fillenham (1963).

Everglades; there, rates of subsidence of 3.2 cm per year have been recorded.

The moisture content of the soil can also affect the degree to which soils are subjected to expansion and contraction effects, which in turn may affect engineering structures in areas with expansive soils. Particular problems are posed by soils containing **smectite** clays. When drained they may dry out and shrink and the soil may crack, damaging the foundations of buildings.

In Britain there has been considerable debate about the effects on river flows, and in particular on flood peaks, of draining upland peat areas for afforestation. There appear to be some cases where flood peaks have increased after peat drainage and others where they have decreased. It has been suggested that differences in peat type alone might account for the different effects. Thus it is possible that the drainage of a catchment dominated by the *Sphagnum* moss would lead to increased flooding, since drainage compacts *Sphagnum*, reducing both its storage volume and its permeability. On the other hand, in the case of peat where *Sphagnum* moss did not grow there would be relatively less change in structure, but there would be a reduction in moisture content and an increase in storage capacity, thereby tending to reduce flood flows. The nature of the peat is, however, just one feature to be considered. The intensity of the drainage works (depth, spacing, etc.) may also be important. In any case, there may be two (sometimes conflicting) processes operating as a result of peat drainage: the increased drainage network will encourage rapid runoff, while the drier soil conditions will provide greater storage for rainfall. Which of these two tendencies is dominant will depend on local catchment conditions.

The impact of land drainage upon the incidence of floods downstream has also long been a source of controversy. This impact depends on the size of the area being considered, the nature of land management and the character of the soil that has been drained. Robinson (1990) conducted a detailed review of experience in the UK and found that the drainage of heavy clay soils that are prone to prolonged surface saturation in their undrained state generally led to a reduction of large and medium flow peaks. He attributed this to the fact that their natural response, with limited soil water storage available, is 'flashy', whereas their drainage largely eliminates surface saturation. By contrast, the drainage of permeable soils, which are less prone to such surface saturation, improves the speed of subsurface flow, thereby tending to increase peak flow levels.

As with so many environmental issues, it is not always easy to determine whether an increase in flood frequency or intensity is the result of land-use changes of the type we have been discussing, or whether some natural changes in rainfall have played a dominant role. In central and southern Wales, for example, there is some clear evidence of changes in the magnitude and frequency of floods over recent decades. This has sometimes been attributed to the increasing amount of afforestation that has been carried out by the Forestry Commission since the First World War, and to the drainage of upland areas that this has necessitated. While in the Severn catchment this appears to be a partial explanation, in other river basins the main cause of more frequent and intense floods appears to have been a marked increase in the magnitude and frequency of heavy daily rainfalls. For example, in the case of the Tawe Valley near Swansea, of 17 major floods since 1875, 14 occurred between 1929 and 1981 and only 3 between 1875 and 1928. Of 22 widespread heavy rainfalls in the Tawe catchment since 1875, only 2 occurred during 1875–1928, but 20 between 1929 and 1981 (Walsh et al., 1982).

FURTHER READING

Robinson, M., 1990, *Impact of Improved Land Drainage on River Flows*. Institute of Hydrology, Wallingford, UK, Report no. 113.
A state-of-the-art review produced by the UK's leading institute for the study of hydrology.

6 WATER POLLUTION

The activities of humans have begun to dominate the quality of natural river waters, both locally and, increasingly, at a regional scale. The ever-increasing human population and its growing wasteload have begun to overtax the recycling capabilities of rivers. The water pollution challenges that the world faces are enormous. They can be categorized, according to source, into three main groups:

- *Municipal waste.* This is composed primarily of human excreta. While it contains relatively few chemical contaminants, it carries numerous pathogenic micro-organisms.
- *Industrial wastes.* These are of very varied composition, depending upon the type of industry or processing activity, and they may contain a wide variety of both organic and inorganic substances.
- *Agricultural wastes.* These are composed of the excess phosphorus and nitrogen present in synthetic fertilizers and in animal wastes, as well as residues from a number of pesticides and herbicides.

It is also possible to categorize water pollutants according to whether or not they are derived from 'point' or 'non-point' (also called 'diffuse') sources (figure IV.6). Municipal and industrial wastes tend to fall into the former category because they are emitted from one specific and identifiable place (e.g. a sewage pipe or industrial outfall). Pollutants from non-

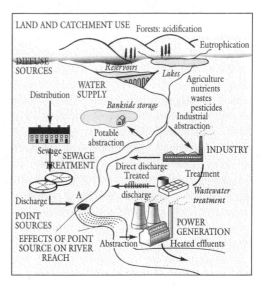

Figure IV.6 Diffuse and point sources of pollution into river systems
Source: After Newson (1992), fig. 7.7.

point sources include agricultural wastes, many of which enter rivers in a diffuse manner as chemicals percolate into groundwater or are washed off into fields, as well as some mining pollutants, uncollected sewage and some urban stormwater runoff.

Possibly the most useful way to categorize pollutants is on the basis of their chemical, physical or biological composition, and this is the framework we shall use for the rest of this section. We will not cover the whole range of waste pollutants, but concentrate on three groups:

- nitrates and phosphates;
- metals;
- synthetic and industrial organic pollutants.

Nitrates and phosphates are an important cause of a process called eutrophication (see section 7 below). Nitrates normally occur in drainage waters and are derived from soil nitrogen, from nitrogen-rich geological deposits and from atmospheric deposition. Anthropogenic sources include synthetic fertilizers, sewage and animal wastes from feedlots. Land-use changes (e.g. logging) can also increase nitrate inputs to streams. Perhaps as much as one-third of the total dissolved nitrogen in river waters throughout the world is the result of pollution. Indeed, Peierls et al. (1991) have demonstrated that the quantity of nitrates in rivers worldwide now appears to be closely linked to the density of human population nearby. Using published data for 42 major rivers, they found a highly significant correlation between nitrate concentration and human population density that explained 76 per cent of the variation in nitrate concentration for the 42 rivers. They maintain that 'human activity clearly dominates nitrate export from land.' Nitrate levels in English rivers are now clearly rising. Current levels (1990s) are between 50 per cent and 400 per cent higher than a quarter of a century ago.

Phosphate levels are also rising in some parts of the world. Major sources include detergents, fertilizers and human wastes.

Metals are another major class of pollutants. Like nitrates and phosphates, metals occur naturally in soil and water. However, as the human use of metals has burgeoned, so has the amount of water pollution they cause. Other factors also contribute to water pollution from metals. Some metal ions reach river waters because they become more quickly mobilized as a result of acid rain (see part III, section 8). Aluminium is a notable example of this. From a human point of view, the metals of greatest concern are probably lead, mercury, arsenic and cadmium, all of which have adverse health effects. Other metals can be toxic to aquatic life, and these include copper, silver, selenium, zinc and chromium.

The anthropogenic sources of metal pollution include the industrial processing of ores and minerals, the use of metals, the leaching of metals from garbage and solid waste dumps, and animal and human excretions. Nriagu and Pacyna (1988) estimated the global anthropogenic inputs of trace metals into aquatic systems (including the oceans), and concluded that the sources producing the greatest quantities were, in descending order, the following (the metals produced by each source are listed in parentheses):

- domestic wastewater effluents (arsenic, chromium, copper, manganese, nickel);
- coal-burning power stations (arsenic, mercury, selenium);
- non-ferrous metal smelters (cadmium, nickel, lead, selenium);
- iron and steel plants (chromium, molybdenum, antimony, zinc);
- the dumping of sewage sludge (arsenic, manganese, lead).

However, in some parts of the world metal pollution may be derived from other sources. There is increasing evidence, for example, that in the western USA water derived from the drainage of irrigated lands may contain high concentrations of toxic or potentially toxic **trace elements** such as arsenic, boron, chromium, molybdenum and selenium. These can cause human health problems and poison fish and wildlife in desert wetlands (Lently, 1994).

Synthetic and industrial organic pollutants have been manufactured and released in very large quantities since the 1960s. The dispersal of these substances into watercourses has resulted in widespread environmental contamination. There are many tens of thousands of synthetic organic compounds currently in use, and many are thought to be hazardous to human health and to aquatic life, even at quite low concentrations – concentrations possibly lower than those that can routinely be measured by commonly available analytical methods. Among these pollutants are synthetic organic pesticides, including chlorinated hydrocarbon insecticides (e.g. DDT). Some of these can reach harmful concentrations as a result of **biological magnification** in the **food chain**. Other important organic pollutants include PCBs, which have been used extensively in the electrical industry as di-electrics in large transformers and capacitors; PAHs, which result from the incomplete burning of fossil fuels; various organic solvents used in industrial and domestic processes; phthalates, which are plasticizers used, for example, in the production of polyvinyl chloride resins; and DBPs, which are a range of disinfection by-products. The long-term health effects of cumulative exposure to such substances are difficult to quantify. However, some work suggests that they may be implicated in the development of birth defects and certain types of cancer.

FURTHER READING

Nash, L., 1993, Water quality and health. In P. H. Gleick (ed.), *Water in Crisis: A Guide to the World's Freshwater Resources*, 25–39. New York: Oxford University Press. An excellent summary of pollution characteristics and effects.

Past and present pollution of the River Clyde, Scotland

The River Clyde, which runs through Glasgow in Scotland, has a mean discharge of 41 cu metres per second. It is tidal in its lower sections, up to the Tidal Weir upstream of the Albert Bridge. It has had a long history of pollution. In 1872 the Royal River Pollution Commission found the Clyde to be the most polluted river in Scotland. Parts were described as a 'foul and stinking flood'. Until the beginning of the nineteenth century the river was probably quite clean, even in the heart of Glasgow. However, by 1845–50 fish populations had been eliminated from the upper estuary. Poor oxygen conditions prevented them from returning until 1972 (McLusky, 1994). In 1872 the Clyde through Glasgow was described thus: 'its water is loaded with sewage mud, fould with sewage gas and poisoned by sewage waste of every kind – from dye works, chemical works, bleach works, paraffin oil works, tanyards, distilleries, privies and water closets' (quoted in Hammerton, 1994).

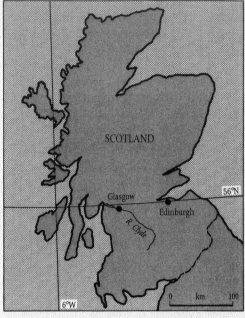

This alarming pollution had come about because of the enormous boom in population and industry in the area. In summer many of the lower tributaries, e.g. the Black Cart and White Cart, were no more than open sewers conveying sewage and industrial wastes to the main river. By the 1930s, over fifty years after the Commission's report, the river was, if anything, in a worse condition. Progress in cleaning it up was hindered by the two world wars, and it was only in 1965 that effective legislation began to improve things. In 1968, when the first biological surveys of the Clyde were done (figure IV.7(a)), no fish were found within the boundary of Glasgow, nor in the lower reaches of the North Calder, South Calder, Kelvin, Black Cart and White Cart. By autumn 1983 Atlantic salmon (*Salmo salmar*) had returned to the Clyde and some fish are now found in all the river areas shown in figure IV.7. Since 1972 dissolved oxygen levels in the Clyde estuary have improved markedly. The greatly improved pollution situation achieved by 1988 is shown in figure IV.7(b). The number of fish species in the upper estuary has steadily increased, to 18 in 1978, 34 in 1984 and 40 in 1992. Thus, even rivers with a long history of dire pollution can be cleaned up and their fauna and flora restored.

Plate IV.6 The River Clyde in Central Glasgow. (Graham Burns/ Environmental Picture Library)

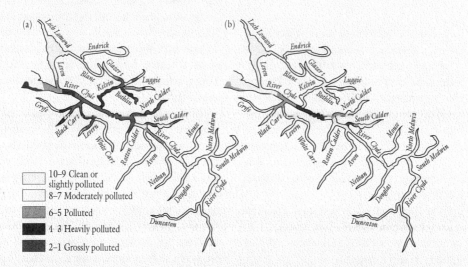

Figure IV.7 The changing pollution of the River Clyde, Scotland, based on biological classification of pollution: (a) 1968; (b) 1988
Source: After Hammerton (1994), figs 20.1, 20.2.

7   EUTROPHICATION

The process of eutrophication can be well illustrated by the case of the Black Sea.

The Black Sea is a very large body of water surrounded by land except for its narrow, shallow connection to the Mediterranean Sea, called the Bosporus. It receives river discharge from a land area five times greater than its own and covering parts of nine different countries. Two of Europe's largest rivers, the Danube and the Dneiper, flow into it. Over 162 million people live within the catchments of these rivers (Mee, 1992). Therefore, pollution generated by all these people heads for the Black Sea. The Danube, for example, currently introduces 60,000 tons of phosphorus per year and some 340,000 tons of total inorganic nitrogen into the Sea.

As a result, and in spite of its size, the Black Sea shows many of the classic symptoms of a process called eutrophication. The symptoms include:

• A gradual shallowing, right across the basin, of the so-called euphotic zone (the surface layer of water in which the light level is sufficient for **net biological primary production**). In other words, the lake is becoming more turbid or cloudy, thereby reducing the amount of light available to support life. The decreased light penetration has resulted in the massive loss of large shallow-water plants.
• Dense **blooms** of a single species of **nanoplankton** have developed, dramatically modifying the base of the marine food chain.
• Widespread **hypoxia** (reduction of oxygen levels) resulting from the enormous increase in organic matter falling to the shelf floor from blooming and decaying organisms. This has led to the complete elimination of a large proportion of **macrobenthic** organisms and the demise of formerly rich commercial fisheries.

What precisely is eutrophication? Fundamentally it is the enrichment of waters by nutrients. Among these nutrients, phosphorus and nitrogen are particularly important as they regulate the growth of aquatic plants. The process does occur naturally – for example, when lakes get older – but it can be accelerated by human activities, both by runoff from fertilized and manured agricultural land and by the discharge of domestic sewage and industrial effluents. This anthropogenically accelerated eutrophication – often called 'cultural eutrophication' – commonly leads, as in the case of the Black Sea, to excessive growths of algae, serious depletion of dissolved oxygen as algae decay after death and, in extreme cases, to an inability to support fish life. It can affect all water bodies, from streams, to lakes, to estuaries and coastal seas. Coastal and estuary waters are sometimes affected by algal foam and scum, often called 'red tides'. Some of these blooms are so toxic that consumers of seafood that has been exposed to them can be affected by diarrhoea, sometimes fatally.

The nature of red tides has recently been discussed by Anderson (1994), who points out that these blooms, produced by certain types of phytoplankton (tiny pigmented plants), can grow in such abundance that they change the colour of the seawater not only to red but also to brown or even green. They may be sufficiently toxic to kill marine animals such as fish and seals. Long-term studies at the local and regional level in many parts of the world suggest that these so-called red tides are increasing in extent and frequency as coastal pollution worsens and nutrient enrichment occurs more often.

Eutrophication also has adverse effects on coral reefs. This has been explained by Weber (1993, p. 49):

Initially, coral productivity increases with rising nutrient supplies. At the same time, however, corals are losing their key advantage over other organisms: their symbiotic self-sufficiency in nutrient-poor seas. As eutrophication progresses, algae start to win out over corals for newly opened spaces on the reef because they grow more rapidly than corals when fertilized. The normally clear waters cloud as phytoplankton begin to multiply, reducing the intensity of the sunlight reaching the corals, further lowering their ability to compete. At a certain point, nutrients in the surrounding waters begin to overfertilize the corals' own zooxanthellae, which multiply to toxic levels inside the polyps. Eutrophication may also lead to black band and white band disease, two deadly coral disorders thought to be caused by algal infections. Through these stages of eutrophication, the health and diversity of reefs declines, potentially leading to death.

The natural process of eutrophication is shown in figure IV.8, using the example of how a lake ages. What has happened, particularly since the Second World War, is that various human actions have speeded up the natural process (figure IV.9). The growth in fertilizer usage in the last five decades has been increasingly rapid. In spite of the increasing costs of energy supplies and hydrocarbons (from which many of the fertilizers are derived) in the 1970s, world fertilizer production has continued to rise inexorably, and fertilizer-derived nitrates reach groundwater and rivers. For example, the mean annual nitrate concentration of the River Thames, which provides most of London's water supply, increased from around 11 mg per litre in 1928 to 35 mg per litre in the 1980s.

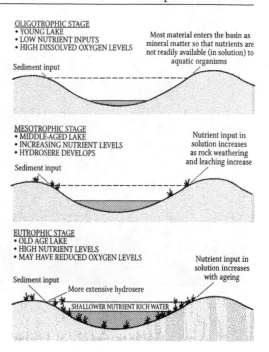

Figure IV.8 The natural process of lake eutrophication
Source: After Mannion (1991), fig. 6.3.

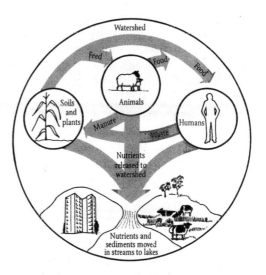

Figure IV.9 The major components of the drainage basin nutrient cycle leading to cultural eutrophication
Source: After Newson (1992), fig. 7.4.

Plate IV.7   Lake Témpé, Sulawesi, Indonesia (Frédéric Pelras).

However, it is necessary to point out that the application of fertilizers is not the only possible cause of rising nitrate levels. Some nitrate pollution may be derived from organic wastes. Intensive cultivation may cause a decline in the amount of organic matter present in the soil, and this could limit a soil's ability to assimilate nitrogen, so that more is lost to water. The pattern of tillage may also affect the liberation of nitrogen. The increased area and depth of modern ploughing accelerates the decay of residues and may change the pattern of water movement in the soil. Finally, the area of England covered by tile drainage has greatly expanded in recent decades. This has affected the movement of water through the soil, accelerating the flow of leached nitrates and other materials into streams.

What can be done to control cultural eutrophication? Preventive measures may include the introduction of laws to limit the type and quantity of permitted discharges from industrial sources. Water companies may be forced to treat effluent to reduce its nutrient content. Bans can be introduced on detergents containing phosphates, as has already been done in some areas. The most severe problems, however, are posed by nutrients derived from agricultural sources. Steps may need to be taken to make agriculture less intensive and to control the application of fertilizers and sludge in locations from which they can easily be washed into streams and rivers, such as floodplains.

FURTHER READING

Mannion, A. M., 1991, *Global Environmental Change*. Harlow: Longman.
This useful general review contains some perceptive information on eutrophication.

Newson, M., 1992, Patterns of freshwater pollution. In M. Newson (ed.), *Managing the Human Impact on the Natural Environment*, 130–49. London: Belhaven Press.
A hydrological approach to understanding the pathways taken by pollutants.

Controlling eutrophication: Lake Biwa, Japan

The largest freshwater lake in Japan, covering 674 sq km, is Lake Biwa, in Shiga Prefecture, Honshu. It is one of the oldest lakes in the world and has a maximum depth of over 100 metres. The catchment area of the lake is less than five times the surface area of the lake itself, and the lake is fed by high annual precipitation and inflow. Until around 1950 Lake Biwa was oligotrophic (i.e. containing low nutrient loads), but since then has become eutrophic (i.e. containing high nutrient loads) with algal blooms first noticed in 1959 and red tides occurring every year since 1977.

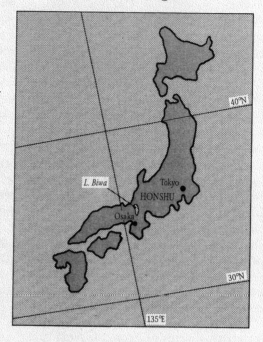

The causes of eutrophic conditions are linked to the explosive economic development of the Shiga Prefecture. Lake Biwa today meets the water needs of 13 million people and all their industry. The lake also provides an important freshwater fishery, and is of immense cultural and spiritual value. Fifty-two per cent of the catchment remains forested, although Japanese red pines have replaced the natural broadleaf forest; 17.3 per cent of the catchment area is now urbanized. The quality of the lake's water has declined as industry and agriculture have expanded and domestic wastes have not been managed effectively (Petts, 1988). Eutrophication peaked in 1978. Since then the lake has shown signs of improvement, as water quality has responded to a number of management strategies.

A ten-year voluntary 'use soap' campaign among local residents reduced dramatically the 18 per cent of the toal phosphorus load that had been coming from domestic detergents. In 1980 a Shiga Prefecture government ordinance regulated industrial, domestic and agricultural discharges of phosphorus and nitrogen. Since then nitrogen and phosphorus concentrations in streams flowing into Lake Biwa have declined by 20 per cent, even though population and industry have continued to grow. Phosphorus levels in the southern part of the lake have also fallen, by 30 per cent. The Shiga Prefecture government has introduced conservation plans to ensure the monitoring of water quality, conservation and environmental education over the long term.

Further reading

Petts, G. E., 1988, Water management: the case of Lake Biwa, Japan. *Geographical Journal* 154, 367–76.

8   THERMAL POLLUTION

Thermal pollution is the pollution of water by increasing its temperature. As many organisms are sensitive to temperature, this form of pollution can have considerable ecological significance.

Where does the heat that produces the thermal pollution come from? One of the main sources in industrialized countries is the condenser cooling water released from power stations. If there are large concentrations of big electricity generating plants along one stretch of river, as for example along the River Trent in the midlands of England, the amount of water involved can be quite large. River water discharged after it has been used for cooling may be some 6–9°C warmer than it was before being taken out of the river. At times of low flow this can raise river water temperatures downstream considerably.

The process of urbanization is another factor that needs to be considered. It has a range of effects: changes produced by the urban heat island effect (see part III, section 5); changes in the temperature of streams brought about by the presence of reservoirs; changes in the volume of storm runoff; and changes in the nature of urban stream channels – how much they are covered over or shaded by vegetation, and how their width and depth compare with natural channels.

Thermal pollution can also occur in rural areas. Large reservoirs will modify downstream river temperatures. Deforestation, which removes shade cover, may increase water temperatures, particularly in the summer months.

Thermal pollution has many ecological effects. Temperature increases can be harmful to temperature-sensitive fish such as trout and salmon and can disrupt spawning and migration patterns (figure IV.10). An increase in water temperature causes a decrease in the solubility of oxygen, which is needed for the oxidation of

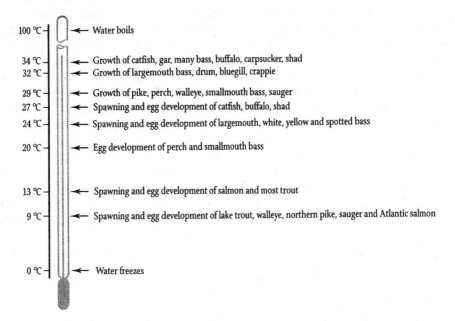

Figure IV.10   Maximum temperatures for the spawning and growth of fish
Source: After Giddings (1975), fig. 13–2.

biodegradable wastes. At the same time, the rate of oxidation is accelerated, demanding more and more oxygen from the smaller supply and thereby depleting the oxygen content of the water still further. Temperature also affects the lower organisms, such as plankton and crustaceans. In general, the higher the temperature is, the less desirable the types of algae in water.

In cooler waters, diatoms are the predominant phytoplankton in water that is not heavily eutrophic; at higher temperatures with the same nutrient levels, green algae begin to become dominant and diatoms decline. At the highest water temperatures, blue-green algae thrive and often develop into heavy blooms.

FURTHER READING

Langford, T. E. L., 1990, *Ecological Effects of Thermal Discharges.* London: Elsevier Applied Science.
The most authoritative advanced treatment of thermal pollution.

9   INTER-BASIN WATER TRANSFERS AND THE DEATH OF THE ARAL SEA

Increasing rates of water consumption and the unequal distribution of water resources from one region to another mean that in many parts of the world long-distance transfers of water are made between river basins. Also, in the world's drylands, large quantities of water are abstracted from rivers to supply irrigation schemes. One of the results of such large-scale modifications of river regimes is that the discharges of some rivers have declined very substantially. This in turn means that the extent and volume of any lakes into which they empty have been reduced.

Perhaps the most severe change to a major inland sea or lake is that taking place to the Aral Sea in the southern part of the former Soviet Union (figure IV.11). Until very recently this was the world's fourth largest lake, with a high level of biological activity and a rich and distinctive aquatic fauna and flora. It had considerable commercial fisheries, and was used for transport as well as sporting and recreational activities. It was also a refuge for huge flocks of waterfowl and migratory birds. It may also have exerted a favourable climatic, hydrological and hydrogeological effect on the surrounding area.

However, since the 1960s a dramatic change has taken place. The inflow of water into the lake has decreased markedly (see figure IV.12), and it has now lost more than 40 per cent of its area and about 60 per cent of its water volume. The lake's level has fallen by more than 14 metres. Its salinity has increased threefold. Its fauna and flora have been destroyed, so that only a small number of aquatic species has survived. The climate around the lake may also have been affected. The increasing areas of exposed, desiccating and salty lake bed provide an ideal environment for the genesis of dust storms. Such storms now evacuate some tens of millions of tons of salt each year and dump them on agricultural land, reducing crop yields. The human population also seems to be suffering from poorer-quality water supply and from respiratory disorders caused by the blowing salt and dust. It is not surprising, therefore, that the Aral Sea is now regarded as the greatest ecological tragedy of the former Soviet Union.

Why has the inflow of water to the Aral Sea declined so extraordinarily? The main reason was that in the 1950s and early 1960s a decision was taken to expand

Plate IV.8   Inter-basin water transfers are vital for the survival of Los Angeles. This large canal transports water from inland California (east of the Sierra Nevada mountains) to satisfy the needs of the sprawling conurbation hundreds of kilometres away. (A. S. Goudie)

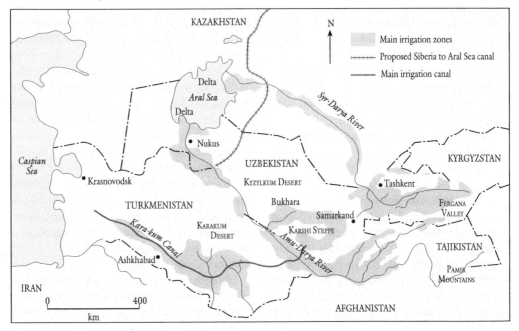

Figure IV.11   Irrigation and the Aral Sea

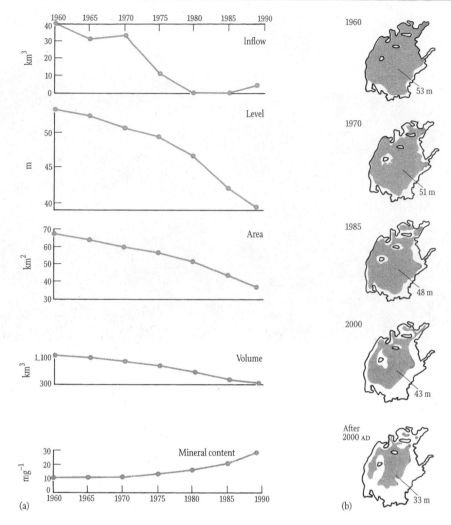

Figure IV.12   (a) Changes in the Aral Sea, 1960–1989; (b) The past and predicted contraction of the Aral Sea as its level falls
Sources: (a) After Kotlyakov (1991); (b) Modified after Hollis (1978), p. 63.

irrigation in Central Asia and Kazakhstan, so that crops like rice and cotton, which consume a great deal of water, could be cultivated in the middle of a desert. Large volumes of fertilizers and herbicides were also used in growing these crops, and these have contributed to the deterioration in water quality. In many cases, too, the irrigation systems themselves were of poor design, construction and operation.

Scientists, economists and politicians are now seeking solutions to the Aral tragedy. Some proposed ideas are:

- The local population should, for health reasons, be provided with supplies of safe, non-polluted, piped water.
- The policy of growing cotton in the deserts of Central Asia needs to be reconsidered.

Plate IV.9 The rapid desiccation of the Aral Sea, following the extraction of water from the rivers that feed it to supply irrigation schemes, has left these boats high, dry and redundant. (Marcus Rose, Panos Pictures)

- Agriculture needs to be made more efficient by reducing the very substantial losses caused by inadequate storage and transport infrastructure.
- A fundamental restructuring of the region's economy should be oriented towards developing products that do not require the intensive use of water.
- Irrigation systems need to be reconstructed to reduce water losses, and application of water should be rationally controlled.
- A policy of expanding the area of irrigated agriculture should be replaced by more intensive use of existing irrigated tracts through better crop rotations, technologies and the growth of improved varieties of crops.
- Attempts should be made to revegetate desiccated areas to reduce dust storm activity.

FURTHER READING

Micklin, P. P., 1988, Desiccation of the Aral Sea: a water management disaster in the Soviet Union. *Science* 241, 1170–5.
One of the key papers that drew attention to the situation around the Aral Sea.

Micklin, P. P., 1992, The Aral crisis: introduction to the special issue. *Post-Soviet Geography* 33 (5), 269–82.
A collection of papers on all aspects of the Aral Sea problem.

10 GROUNDWATER DEPLETION AND GROUNDWATER RISE

In some parts of the world groundwater is the main source of water for industrial, municipal and agricultural use. Some rocks, including sandstones and limestones, have characteristics that enable them to hold and transmit large quantities of water, which can be reached by installing pumps and boreholes. In some cases large amounts of water can be abstracted without severe environmental effects. This is particularly true of areas where a combination of favourable climate, topography, geological structure and rock composition enables the water-bearing body – the aquifer – to recharge itself rapidly. In such cases the water is a renewable resource. However, in other cases the rate of exploitation may greatly exceed the capacity of the aquifer to be recharged. In these cases both the quantity and the quality of the water supply may deteriorate markedly over time. In such cases it is more appropriate to refer to the extraction of water as 'mining' of a largely non-renewable resource.

Let us take an extreme example: the exploitation of groundwater resources in the oil-rich kingdom of Saudi Arabia. Most of Saudi Arabia is desert, so that climatic conditions are not favourable for rapid large-scale recharge of aquifers. Also, much of the groundwater that lies beneath the desert is a fossil resource, created during more humid conditions – pluvials – that existed in the Late Pleistocene, between 15,000 and 30,000 years ago. In spite of these inherently unfavourable circumstances, Saudi Arabia's demand for water is growing inexorably as its economy develops. In 1980 the annual demand was 2.4 billion cubic metres (bcm). By 1990 it had reached 12 bcm (a fivefold increase in just a decade), and it is expected to reach 20 bcm by 2010. Only a very small part of the demand can be met from desalination plants or surface runoff; over three-quarters of the supply is obtained from predominantly non-renewable groundwater resources. The **drawdown** on aquifers is thus enormous. It has been calculated that by 2010 the deep aquifers will contain 42 per cent less water than in 1985. Much of the water is used ineffectively and inefficiently in the agricultural sector (Al-Ibrahim, 1991), to irrigate crops that could easily be grown in more humid regions and then imported.

Saudi Arabia is not alone in its voracious appetite for groundwater. In many parts of the world such problems have grown with increasing population levels and consumption demands, together with the adoption of new exploitation techniques (for example, the replacement of irrigation methods involving animal or human power by electric and diesel pumps).

Considerable reductions in groundwater levels have been caused by abstraction in other areas. The rapid increase in the number of wells tapping groundwater in the London area from 1850 until after the Second World War caused substantial changes in groundwater conditions. The **piezometric surface** in the confined chalk aquifer has fallen by more than 60 metres over hundreds of square kilometres. Likewise, beneath Chicago, Illinois, USA, pumping since the late nineteenth century has lowered the piezometric head by some 200 metres. The drawdown that has taken place in the Great Artesian Basin of Australia exceeds 80–100 metres in some places. The environmental consequences of excessive groundwater abstraction include salinization of coastal aquifers (see part V, section 5) and land subsidence (see part V, section 7).

Some of the most serious reductions in water levels are taking place in the High

Plate IV.10   A major cause of groundwater depletion is the use of centre-pivot irrigation schemes. The Ogallala aquifer of the High Plains of Texas, where this photo was taken, is a major example of this phenomenon. (A. S. Goudie)

Plains of Texas, threatening the long-term viability of irrigated agriculture in that area. Before irrigation development started in the 1930s, the High Plains groundwater system was in a state of dynamic equilibrium, with long-term recharge equal to long-term discharge. However, the groundwater is now being mined at a rapid rate to supply **centre-pivot** and other schemes. In a matter of only fifty years or less, the water level has declined by 30–50 metres in a large area to the north of Lubbock, Texas. The aquifer has narrowed by more than 50 per cent in large parts of certain counties, and the area irrigated by each well is contracting as well yields are falling.

In some industrial areas, recent reductions in industrial activity have led to less groundwater being taken out of the ground. As a consequence, groundwater levels in such areas have begun to rise, a trend exacerbated by considerable leakage from ancient, deteriorating pipe and sewer systems. This is already happening in British cities including London, Liverpool and Birmingham. In London, because of a 46 per cent reduction in groundwater abstraction, the water table in the Chalk and Tertiary beds has risen by as much as 20 metres. Such a rise has numerous implications, both good and bad:

- increase in spring and river flows;
- re-emergence of flow from 'dry springs';
- surface water flooding;
- pollution of surface waters and spread of underground pollution;
- flooding of basements;
- increased leakage into tunnels;
- reduction in stability of slopes and retaining walls;

- reduction in bearing capacity of foundations and piles;
- increased **hydrostatic uplift** and swelling pressures on foundations and structures;
- swelling of clays as they absorb water;
- chemical attack on building foundations.

There are various methods of recharging groundwater resources, providing that sufficient surface water is available. Where the materials containing the aquifer are permeable (as in some alluvial fans, coastal sand dunes or glacial deposits) the technique of water-spreading is much used. In relatively flat areas river water may be diverted to spread evenly over the ground so that infiltration takes place. Alternative water-spreading methods may involve releasing water into basins which are formed by excavation, or by the construction of dikes or small dams. On alluvial plains water can also be encouraged to percolate down to the water table by distributing it into a series of ditches or furrows. In some situations natural channel infiltration can be promoted by building small check dams down a stream course. In irrigated areas surplus water can be spread by irrigating with excess water during the dormant season. In sediments with impermeable layers such water-spreading techniques are not effective, and the appropriate method may then be to pump water into deep pits or into wells. This last technique is used on the heavily settled coastal plain of Israel, both to replenish the groundwater reservoirs when surplus irrigation water is available, and to attempt to diminish the problems associated with salt-water intrusion from the Mediterranean.

FURTHER READING

Downing, R. A. and Wilkinson, W. B. (eds), 1991, *Applied Groundwater Hydrology: A British Perspective*. Oxford: Clarendon Press.
An advanced textbook on all aspects of groundwater in the British context.

11 CONCLUSION

Freshwater resources are of vital importance. Their quality and quantity have undergone major changes as a consequence of human activities. Human demand for freshwater has grown inexorably in recent decades. As a result, ever-increasing proportions of river flow are being controlled or modified by deliberate human manipulation, most notably by the construction of dams, the channelization of streams and by long-distance inter-basin water transfers. As our case study of the Colorado River in the USA shows, the degree of control that can be achieved is radical.

Equally important are changes in the hydrological cycle resulting unintentionally from changes in land use and land cover. The replacement of forests with farms, and of countryside with cities, are two particularly important mechanisms in this respect. Also significant are the consequences – some anticipated, some not – of land drainage activities.

Humans have modified not only the quantity of river flow, but also its quality. Many water resources are polluted by a range of both 'point' and 'non-point' sources. River pollution can lead in turn to eutrophication of lakes and inland and marginal seas. However, water pollution and lake eutrophication, as our case studies of the River Clyde and Lake Biwa show, are reversible if proper management strategies are adopted. Nonetheless, as is shown in the case of the rapid desiccation of the Aral Sea, the required solutions may be extremely complex, and technological change is seldom the only solution. Wholesale changes in a region's economic and political structure may be required.

Finally, we have pointed to the importance of groundwater reserves and showed that in some countries these resources are being exploited at an unsustainable rate. In many areas the water table has been drawn down too far and too fast. We have also pointed out that in other parts of the world the reverse process is happening and that groundwater levels are rising.

The issues discussed in this part of the book indicate how problems of human impacts on water are complicated by the links between bodies of water, by the mutual interaction of different stresses, and by the links between water and other aspects of the environment, such as the atmosphere, biosphere and land surface.

KEY TERMS AND CONCEPTS

aquifer
biological magnification
channelization
clear-water erosion
eutrophication

inter-basin water transfers
point and non-point sources of pollution
red tides
thermal pollution

POINTS FOR REVIEW

Are dams a good or a bad thing?

If you were in charge of providing large quantities of clean water in an area, would you seek to increase or decrease the amount of forest in your catchments?

How do humans increase the risk of river floods?

What is the difference between 'point' and 'non-point' sources of water pollution?

What do you understand by the term 'eutrophication'?

How do humans change the state of groundwater bodies?

PART V

The Land Surface

1   INTRODUCTION

This part of the book focuses on human impacts on the land surface – primarily soils and landforms. Humans have modified soils in many ways. Most serious of all has been the impact of land-use changes on the rates at which soils are eroded by wind and water. The quality of soils has also been transformed. At present, for example, many soils in irrigated regions are being affected by salinization; but at other times and in other places changes in soil quality have included the formation of peat layers, **podzols** and laterite hardpans. Soil conditions are affected by modern farming practices: heavy farm machinery causes soil compaction, and chemical changes are brought about by the application of synthetic fertilizers. The whole impact of humans on soils is often termed **metapedogenesis** (table V.1).

As well as soils, the skin of the earth is composed of its landforms. Here again the human impact can be considerable (table V.2). In particular, humans can destabilize hillside slopes and cause wholesale ground subsidence. The range of human impacts on landforms and landforming processes is considerable. Some landforms may be produced by direct anthropogenic processes. Examples are landforms produced by constructional activity (e.g. tipping), excavation, mining and farming (e.g. terracing). Landforms produced indirectly by human activities, while less easy to recognize, are of particular importance. Indeed, the indirect and unintentional modification of processes by humans is the most crucial aspect of what is called **anthropogeomorphology**. The geomorphological effects of removing vegetation are an example of this type of change. Sometimes humans deliberately try to change landforms and landforming processes, but set in train a series of events which were not anticipated or desired. As is noted in part VI, section 3, for instance, many attempts have been made to reduce coastal erosion by building impressive-looking and expensive engineering structures which have in fact exacerbated the erosion rather than halting it.

2   SOIL EROSION BY WATER

Soil erosion is a natural geomorphological process which takes place on many land surfaces. Under grassland or woodland it takes place slowly and appears to be more or less balanced by soil formation. Accelerated soil erosion takes place where humans have interfered with this balance by modifying or removing the natural vegetation cover. Construction, urbanization, war, mining and other such activities are often significant in accelerating the problem. However, the main causes of soil erosion are deforestation and agriculture.

Forests protect the underlying soil from the direct effects of rainfall, generating an environment in which rates of erosion tend to be low. The canopy plays an important role both by shortening the fall of raindrops, and by decreasing the speed at which they hit the ground. There are examples of certain types of trees (e.g. beech) in certain environments (e.g. maritime temperate) creating large raindrops, but in general most canopies reduce the erosive effects of rainfalls. The presence of humus in forest soils may be even more important than the canopy in reducing erosion rates in forest. Humus in the soil both absorbs the impact of raindrops and leads to soils with extremely high permeability. Thus forest soils have high infiltration capacities. Another reason why forest soils allow large quantities of water to pass through them is that they have many large **macropores** produced by roots and their rich soil fauna. Forest soils are also well aggregated, making them resistant to both the effects of wetting and water drop

---

Table V.1   Metapedogenesis: human impacts on the main factors involved in soil formation

Factor	*Human impacts*
Parent material	*Beneficial*: adding mineral fertilizers; accumulating shells and bones; accumulating ash; removing excess amounts of substances such as salts *Detrimental*: removing through harvest more plants and animal nutrients than are replaced; adding materials in amounts toxic to plants or animals; altering soil constituents in a way which depresses plant growth
Topography	*Beneficial*: checking erosion through surface roughening, land forming and structure building; raising land level by accumulation of material; land levelling *Detrimental*: causing subsidence by draining wetlands and by mining; accelerating erosion; excavating
Climate	*Beneficial*: adding water by irrigation; rainmaking by seeding clouds; removing water by drainage; diverting winds *Detrimental*: subjecting soil to excessive insolation, to extended frost action, or to wind and rain
Organisms	*Beneficial*: introducing and controlling populations of plants and animals; adding organic matter; loosening soil by ploughing to admit more oxygen; fallowing; removing pathogenic organisms, e.g. by controlled burning *Detrimental*: removing plants and animals; reducing organic content of soil through burning, ploughing, overgrazing, harvesting; adding or encouraging growth of pathogenic organisms; adding radioactive substances
Time	*Beneficial*: rejuvenating soil by adding fresh parent material or through exposure of local parent material by soil erosion; reclaiming land from under water *Detrimental*: degrading soil by accelerated removal of nutrients from soil and vegetation cover; burying soil under solid fill or water

Source: Modified from Bidwell and Hole (1965).

impact. This high degree of aggregation is a result of the presence of considerable quantities of organic material, which is an important cementing agent in the formation of large, water-stable aggregates. Earthworms also help to produce well-aggregated soil. Finally, deep-rooted trees help to stabilize steep slopes by increasing the total **shear strength** of the soils.

It is therefore to be expected that with the removal of forest, for agriculture or for other reasons, rates of soil loss will rise and mass movements (landslides, debris flows, etc.) will happen more often and on a larger scale. Rates of erosion will be particularly high if the deforested ground is left bare; under crops the increase will be less marked. The method of ploughing,

Table V.2   Major anthropogeomorphological processes

Type of process	Examples
Direct anthropogenic processes	
Constructional	Tipping, moulding, ploughing, terracing
Excavational	Digging, cutting, mining, blasting of cohesive or non-cohesive materials, cratering, tramping and churning
Hydrological interference	Flooding, damming, canal construction, dredging, channel modification, draining, coastal protection
Indirect anthropogenic processes	
Acceleration of erosion and sedimentation	Agricultural activity and vegetation clearance, engineering (especially road construction and urbanization)
	Incidental modifications of hydrological regime
Subsidence	Collapse and settling related to mining, groundwater pumping and permafrost melting (thermokarst)
Slope failure	Landsliding, flow and accelerated creep caused by loading, undercutting, shaking and lubrication
Earthquake generation	Loading by reservoirs, lubrication along fault planes

Source: After Goudie (1993).

the time of planting, the nature of the crop, and the size of the fields will all have an influence on the severity of erosion.

Many fires are started by humans, either deliberately or accidentally. Because fires remove vegetation and expose the ground, they also tend to increase rates of soil erosion. The burning of forests, for example, can lead to high rates of soil loss, especially in the first years after the fire. Rates of soil loss in burnt forests are often up to ten times higher than those in protected areas.

Soil erosion can also be caused by construction and urbanization. A number of studies have been done which illustrate clearly that urbanization can create significant changes in erosion rates. The highest rates of erosion are produced in the construction phase, when there is a large amount of exposed ground and a lot of disturbance from vehicle movements and excavations. The equivalent of many decades of natural or even agricultural erosion may take place during a single year in areas cleared for construction. However, construction does not go on for ever, and eventually the building work is completed. Then the disturbance ceases, roads are surfaced, and gardens and lawns are cultivated. Rates of erosion fall dramatically, perhaps to the levels prevailing under natural or pre-agricultural conditions.

Soil erosion by water takes on a variety of forms. **Splash erosion** and **sheet flow** may occur in some areas. Elsewhere **rills**

Plate V.1   Soil erosion at St Michael's Mission in central Zimbabwe. A large *donga* (or erosional gully) has formed as a result of overgrazing and other land-use pressures. (A. S. Goudie)

(small channels) may develop. Under more extreme conditions, for example where soils are highly erodible, large gullies may form, and these may coalesce to form a **badlands** topography. Slopes can become destabilized so that mass movements occur.

Concern about accelerated erosion focuses on two main categories of impact. The first of these relates to the threat it poses to our ability to grow crops and to feed the world's growing population. Soil erosion reduces soil depth and often means that the most fertile, humus- and nutrient-rich portion of the soil profile is lost. The second category of impact is what are termed 'off-farm impacts'. These include:

- accelerated siltation of reservoirs, rivers, drainage ditches, etc.
- eutrophication of water bodies by the transport of nutrients attached to soil particles;
- damage to property by soil-laden water and debris flows.

There is some evidence that soil erosion is becoming a more serious problem in parts of Britain, in spite of the fact that the country's rainfall is much less intense, and so less erosive, than in many parts of the world. The following practices may have caused this state of affairs:

- Ploughing on steep slopes that were formerly under grass, in order to increase the area of arable cultivation.
- Use of larger and heavier agricultural machinery, which tends to increase soil compaction.
- Use of more powerful machinery which permits cultivation in the direction of

maximum slope rather than along the contour. Rills often develop along the wheel ruts ('wheelings') left by tractors and farm implements, and along drill lines.

- Use of powered harrows in seedbed preparation and the rolling of fields after drilling.
- Removal of hedgerows and the associated increase in field size. Larger fields cause an increase in slope length and thus a higher risk of erosion.
- Declining levels of organic matter resulting from intensive cultivation and reliance on chemical fertilizers, which in turn lead to reduced aggregate stability.
- Widespread introduction of autumn-sown cereals to replace spring-sown cereals. Because of their longer growing season, autumn-sown cereals produce greater yields and are therefore more profitable. The change means that seedbeds with a fine tilth and little vegetation cover are exposed throughout the period of winter rainfall.

Several measures can be used to reduce the rate at which soil is lost from agricultural land. In some parts of the world these techniques have been practised for some time and have been quite successful. They are:

- Revegetation:
  (a) deliberate planting;
  (b) suppression of fire, grazing, etc., to allow regeneration.
- Measures to stop stream bank erosion (e.g. stone banks and **rip-rap**).
- Measures to stop gully enlargement:
  (a) planting of trailing plants, etc.;
  (b) weirs, dams, **gabions**, etc.;
- Crop management:
  (a) maintaining cover at critical times of year;
  (b) rotation of crops;
  (c) growing cover crops;
  (d) agroforestry.
- Slope runoff control:
  (a) terracing;
  (b) deep tillage and application of humus;
  (c) digging transverse hillside ditches to interrupt runoff;
  (d) contour ploughing;
  (e) preservation of vegetation strips (to limit field width).
- Prevention of erosion from point sources such as roads and feedlots:
  (a) intelligent geomorphic location of roads, feedlots, etc.;
  (b) channelling of drainage water to non-susceptible areas;
  (c) covering of banks, cuttings, etc., with vegetation.

Further Reading

Boardman, J., Foster, I. D. L. and Dearing, J. A. (eds), 1990, *Soil Erosion on Agricultural Land*. Chichester: Wiley.
An edited series of advanced research papers providing some useful case studies.

Hudson, N., 1971, *Soil Conservation*. London: Batsford.
A general introductory level textbook.

Morgan, R. P. C., 1995, *Soil Erosion and Conservation*, 2nd edn. Harlow: Longman.
A general introduction that is especially strong on methods of controlling erosion.

Pimental, D. (ed.), 1993, *World Soil Erosion and Conservation*. Cambridge: Cambridge University Press.
A series of advanced edited papers that look at soil erosion in a regional context.

Soil erosion on the South Downs, southern England

The South Downs are a range of chalk hills in south-east England which rise to an altitude of around 200 metres. They are deeply dissected by a network of dry valleys. In the early Holocene the Downs were wooded and their soils were much thicker than they are now. Soils are now typically shallow, and stony **rendzinas** with A horizons usually less than 25 cm thick. Since the Second World War the dominant land use in the area has been farming of wheat and barley. In the 1970s a major change of farming practice occurred with the adoption of autumn-grown cereals (e.g. 'winter wheat') in preference to lower-yielding spring-sown varieties. Farming has also become more intensive: fields have increased in size with the removal of hedges and grass banks, while larger and more powerful tractors have enabled farmers to cultivate slopes as steep as 25°.

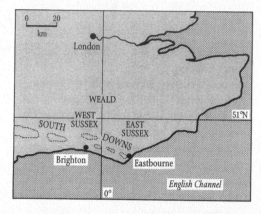

Plate V.2   Soil erosion and flood runoff on the South Downs, south-east England. (John Boardman)

As a result, there has been an increase in soil erosion by water on the Downs, especially between September and December on land prepared for, or drilled with, winter cereals. This is because large areas of smooth ground, with minimal vegetation cover, are exposed during the wettest months. Rills develop in hillsides (figure V.1), especially along tractor wheel ruts ('wheelings'), and some gullies have developed along valley bottoms. Sediment-laden runoff can cause serious problems for nearby houses.

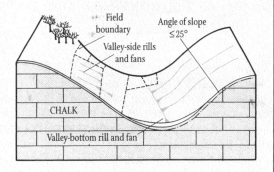

Figure V.1 Typical location and form of erosion on agricultural land on the South Downs, England
Source: After Boardman (1992), fig. 2.1.

Localized erosion and flooding were recorded in 1958 and 1976, but there are few records of such events earlier in the twentieth century. Frequent and sometimes serious erosion occurred in the 1980s, especially during the wet autumns and winters of 1982–3, 1987–8 and 1990–1. John Boardman has monitored about 36 sq km of agricultural land in the area during this time (see table V.3) and found that peak soil erosion in the 1987–8 winter season was accompanied by serious flooding of roads and properties.

Further reading

Boardman, J., 1995, Damage to property by runoff from agricultural land, South Downs, southern England, 1976–1993. *Geographical Journal* 161, 177–91.

Table V.3   Rainfall and soil erosion on a monitored site in the eastern South Downs, England, 1982–1991

Year	Total rainfall, 1 Sep.–1 Mar. (mm)	Total soil loss (cu metres)
1982–3	724	1,816
1983–4	560	27
1984–5	580	182
1985–6	453	541
1986–7	503	211
1987–8	739	13,529
1988–9	324	2
1989–90	621	940
1990–1	469	1,527
1991–2	298	112

Source: Modified from Boardman (1995).

3   WIND EROSION AND DUNE REACTIVATION

In the drier parts of the world, or on particularly light soils, wind erosion may become a major cause of accelerated soil loss. As in the case of accelerated soil erosion by water, the key factor is the removal of vegetation cover.

Possibly the most famous case of soil erosion by deflation was the dust bowl of the 1930s in the USA. This was caused in part by a series of hot, dry years which depleted the vegetation cover and made the soils dry enough to be susceptible to wind erosion. The effects of this drought were made very much worse by years of overgrazing and unsatisfactory farming techniques. However, perhaps the primary cause of the dust bowl was the rapid expansion of wheat cultivation in the Great Plains. The number of cultivated hectares doubled during the First World War as tractors (for the first time) rolled out on to the plains in their thousands. In Kansas alone wheat cultivation increased from under 2 million hectares in 1910 to almost 5 million in 1919. After the war wheat cultivation continued apace, helped by the development of the combine harvester and government assistance. The farmer, busy sowing wheat and reaping gold, could foresee no end to his land of milk and honey; but the years of favourable climate were not to last, and over large areas the tough sod which exasperated the earlier homesteaders had given way to friable soils which were very susceptible to erosion. Drought, acting on damaged soils, created the 'black blizzards' that carried dust as far as the Atlantic seaboard (see box in part III, section 2).

The dust bowl was not solely a feature of the 1930s, and dust storms are still a serious problem in various parts of the United States. For example, in the San Joaquin Valley area of central California in 1977 a dust storm caused extensive damage and erosion over an area of about 2,000 sq km. More than 25 million tonnes of soil were stripped from grazing land within a 24-hour period. While the combination of drought and a very high wind (as much as 300 km per hour) provided the predisposing conditions for the stripping to occur, overgrazing and the general lack of windbreaks in the agricultural land played a more significant role. In addition, broad areas of land had recently been stripped of vegetation, levelled or ploughed up prior to planting. Other contributory factors, albeit quantitatively less important, included the stripping of vegetation for urban expansion, extensive denudation of land in the vicinity of oilfields, and local denudation of land by recreational use of vehicles. One interesting observation made in the months after the dust storm was that, in subsequent rainstorms, runoff occurred faster from those areas that had been stripped by the wind, exacerbating problems of flooding and creating numerous gullies. Elsewhere in California dust yield has been considerably increased by mining operations in dry lake beds.

A comparable acceleration of dust storm activity has also occurred in the former Soviet Union. After the 'Virgin Lands' programme of agricultural expansion in the 1950s, dust storms in the southern Omsk region became on average two and a half times more frequent, and in some local areas five or six times more frequent.

We can see how drought and human pressures can combine to produce accelerated wind erosion by considering the meteorological data for dust-storm frequency at Nouakchott in Mauritania, western Africa (figure V.2). Since the 1960s the number of dust storms has gone up dramatically from just a few each year to over 80 a year. This is partly caused by the

Plate V.3 Gully erosion by water near Luyengo, central Swaziland, southern Africa. (A. S. Goudie)

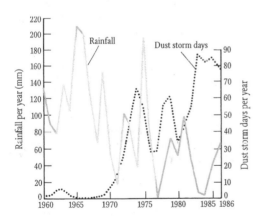

Figure V.2 Annual frequency of dust storm days and annual rainfall for Nouakchott, Mauritania, Africa, 1960–1986

Source: After Goudie and Middleton (1992), fig. 8.

low rainfall of the Sahel drought that has afflicted the area, but also by human population growth and increased disturbance of the desert surface by vehicles.

Wind erosion is also active in certain parts of Britain. Dust storms have been recorded in the Fenlands, the Brecklands, East Yorkshire and Lincolnshire since the 1920s, and they seem to be happening more often in recent years. The storms result from changing agricultural practices, including the use of artificial fertilizers in place of farmyard manure, a reduction in the process of 'claying', whereby clay was added to the peat to stabilize it, the removal of hedgerows to facilitate the use of bigger farm machinery, and, perhaps most importantly, the increased cultivation of sugar beet. This crop requires a fine tilth and tends to leave the soil relatively bare in early summer compared with other crops.

When and where wind erosion of soils takes place is determined by two sets of factors: wind erosivity and surface erodibility (table V.4). Wind erosion will normally be reduced if those variables marked in the table with a (+) are increased and if those marked with a (−) are reduced. (Those marked with a ± can have either effect.) These are important points to be considered when selecting conservation methods, which have to focus on improving the stability of the soil and reducing wind velocities at the soil surface.

Soil conservation measures can be divided into three types:

- *agronomic measures*, which manipulate vegetation to minimize erosion by protecting the soil;
- *soil management methods*, which focus on ways of preparing the soil to promote good vegetative growth and improve soil structure in order to increase resistance to erosion;
- *mechanical methods*, which manipulate the surface topography in order to reduce wind velocity and turbulence.

Agronomic measures use living vegetation or the residues from harvested crops to protect the soil. Wind erosion problems occur on croplands only when the soil is exposed because the crop is not mature enough to provide adequate protection. Hence 'stubble mulching', which involves tilling but not to the extent whereby the field is left 'clean', has become a widely used method of protection from erosion. Soil management techniques are concerned with different methods of soil tillage, the farmer's methods of preparing a suitable seedbed for crop growth, and of helping to control weeds. Mechanical methods include the creation of protective barriers against the wind, such as fences, windbreaks and shelter belts.

Another aspect of wind erosion is dune reactivation. This occurs on the margins of the great subtropical and tropical deserts and is one facet of the process of desertification (see part II, section 3). Dune reactivation arouses some of the strongest fears among those combating desertification. The increasing population levels of both humans and their domestic animals, brought about by improvements in health and by the provision of boreholes, has led to excessive pressure on the limited vegetation resources. As ground cover has been reduced, so dune instability has

Table V.4   Key factors influencing wind erosion of soils[a]

Erosivity	Wind variables	
	Velocity	−
	Frequency	−
	Duration	−
	Magnitude	−
	Shear	−
	Turbulence	−
Erodibility	Debris variables	
	Particle size	±
	Soil clods and cohesive properties	+
	Abradability	−
	Transportability	−
	Organic matter	+
	Surface variables	
	Vegetation: residue	+
	height	+
	orientation	+
	density	+
	fineness	+
	cover	+
	Soil moisture	+
	Surface roughness	+
	Surface length (distance from shelter)	−
	Surface slope	±

[a] See text above for explanation.
Source: After Cooke and Doornkamp (1993).

Plate V.4   The use of palm frond fences to reduce sand movement at Erfoud, southern Morocco. (A. S. Goudie)

increased. The problem is not so much that dunes in the desert cores are relentlessly marching on to moister areas, more that fossil dunes, laid down during the more arid phase peaking around 18,000 years ago, have been reactivated *in situ* by the removal of stabilizing vegetation.

Many methods are used in the attempt to control drifting sand and moving dunes. In practice, most solutions to the problem of dune instability and sand blowing have involved establishing a vegetation cover. This is not always easy. Plant species used to control sand dunes must be able to endure undermining of their roots, burying, abrasion and often severe deficiencies of soil moisture. Thus the species selected need to have the ability to recover after partial burying, to have deep and spreading roots, to have rapid height growth in the seedling stages, to promote rapid litter development, and to add nitrogen to the soil through root nodules. During the early stages of growth they may need to be protected by fences, sand traps and surface mulches. Growth can also be stimulated by the addition of synthetic fertilizers.

FURTHER READING

Goudie, A. S. (ed.), 1990, *Techniques for Desert Reclamation*. Chichester: Wiley. This edited work contains several chapters on the control of dunes and dust hazards.

Controlling sand at Walvis Bay, Namibia

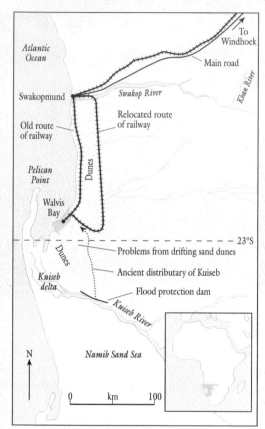

Figure V.3   The coastal zone near Walvis Bay, Namibia. The main coast road is often blocked by sand, while the main railway line has had to be relocated inland

The movement of sand can be a serious problems for the inhabitants of the world's drylands. Sand storms reduce visibility on roads, while encroaching dunes can overwhelm houses, farms, canals and transport links. For this reason, humans have developed a range of techniques to try to control drifting sand and moving dunes (table V.5).

One location where moving sand has proved to be a severe problem, and where many of these control techniques have been attempted, is the port town of Walvis Bay on the coast of Namibia in southern Africa (figure V.3). Here the annual rainfall is so low (around 25 mm) that vegetation cover is minimal, allowing sand to move when wind velocities reach critical levels (as they often do). Many of the dunes near Walvis Bay are individual crescent-shaped dunes called barchans. These are highly mobile, travelling some tens of metres per year. From time to time they have blocked roads; they have also caused the relocation of the main railway link with the interior. The local authority is trying to stabilize the sand by such means as planting (and irrigating) shrubs, and building sand fences (see plates V.4, V5).

Plate V.5 A road has been blocked by a migrating barchan dune near Walvis Bay, Namibia. Sand fences have been constructed in an attempt to slow the process down. (H. A. Viles)

Table V.5   Control techniques for drifting sand and mobile dunes

Problem	Control methods
Drifting sand	Enhancement of deposition of sand through creating large ditches, vegetation belts and barriers, and fences Enhancement of sand transport by aerodynamic streamlining of surface, or changing surface materials Reduction of sand supply by surface treatment, improved vegetation cover or erection of fences Deflection of moving sand by fences, barriers, or vegetation belts
Moving dunes	Removal by mechanical excavation Destruction by reshaping, trenching through dune axis or surface stabilization of barchan arms Immobilization by trimming, surface treatment and fences

4   River Channel Changes

River channels developed in alluvium (sediments deposited by the river itself) adjust their shape, slope and velocity of flow in response to discharge, sediment load, the calibre of the bed and bank sediment, the bank's vegetation and the slope of the valley. Humans have intervened in natural channel systems by building dams, realigning channel courses, constructing levées and embankments, and in many other ways (see part IV, section 2). However, they have also altered channel characteristics unintentionally (table V.6), for example by modifying the amounts of discharge, the amount of sediment being carried, and the nature of vegetation on the river bank. Sometimes, too, deliberate changes have set in train a series of unintended changes.

Let us consider channel straightening. For purposes of navigation and flood control, humans have deliberately straightened many river channels. The elimination of **meanders** contributes to flood control in two ways. First, it eliminates some floods over banks on the outside of curves, where the current is swiftest and where the water

surface rises highest. Second, and more importantly, the shorter straightened course increases both the gradient and the velocity of the stream. Floodwaters can then erode and deepen the channel, thereby increasing its flood capacity. Deliberate channel-straightening causes various types of adjustment in the channel, both within and downstream from straightened reaches. The types of adjustment vary according to such influences as stream gradient and sediment characteristics (figure V.4). Brookes (1988) has recognized five types of change within the straightened reaches (types W1 to W5) and two types of change downstream (types D1 and D2):

- Type W1 is degradation of the channel bed. This happens because straightening shortens the channel path and therefore increases the slope. This in turn increases the efficiency of sediment transport.
- Type W2 is the development of an armoured layer on the channel bed by the more efficient removal of fine materials as described under Type W1.
- Type W3 is the development of a sinuous **thalweg** in streams which are not

Table V.6   Accidental channel changes

Phenomenon	Cause
Channel incision	'Clear-water erosion' below dams caused by sediment removal
Channel aggradation	Reduction in peak flows below dams Addition of sediment to streams by mining, agriculture, etc.
Channel enlargement	Increase in discharge level produced by urbanization
Channel diminution	Discharge decrease following water abstraction or flood control
Channel diminution	Trapping and stabilizing of sediment by artificially introduced plants

Plate V.6 *A completely artificial stream channel in Maspalomas, Gran Canaria, Canary Islands. (A. S. Goudie)*

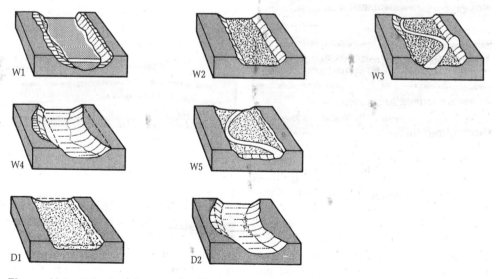

Figure V.4 Principal types of adjustment in straightened river channels
Source: After Brookes (1987), fig. 4.

only straightened but also widened beyond the width of the natural channel.

- Type W4 is the recovery of sinuosity as a result of bank erosion in channels with high slope gradients.
- Type W5 is the development of a sinuous course by deposition in streams with a high sediment load and a relatively low valley gradient.
- Types D1 and D2 result from deposition downstream as the stream tries to even out its gradient. The deposition may occur either as a general raising of the level of the bed, or as a series of accentuated point bar deposits.

Another influence on channel form is the growth of settlements. It is now widely recognized that the urbanization of a river basin results in an increase in the peak flood flows in a river (see part IV, section 4). It is also recognized that the morphology of stream channels is related to their discharge characteristics, and especially to the discharge at which bank full flow, that is, a complete filling of the channel, occurs. As a result of urbanization, the frequency of discharges which fill the channel will increase. This will mean that the beds and banks of channels in erodible materials will be eroded so as to enlarge the channel. This in turn will lead to bank caving, possible undermining of structures, and increases in turbidity.

Similarly important changes in channel morphology result from the lowering of discharge caused by flood-control works and diversions for irrigation. This can be shown for the North Platte and the South Platte Rivers in America, where both peak discharge and mean annual discharge have declined to 10–30 per cent of their pre-dam values. The North Platte, 762–1,219 metres wide in 1890 near the Wyoming–Nebraska border, has narrowed to about 60 metres at present. The South Platte River 89 km above its junction with the

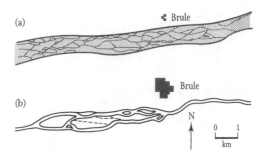

Figure V.5   The configuration of the channel of the South Platte River at Brule in Nebraska, USA, (a) in 1897 and (b) in 1959. Such changes in channel form result from discharge diminution caused by flood control works and diversions for irrigation
Source: After Goudie (1993), fig. 6.15(a).

North Platte was about 792 metres wide in 1897, but had narrowed to about 60 metres by 1959. The tendency of both rivers has been to form one narrow, well-defined channel in place of the previously wide, braided channels. The new channel is also generally somewhat more sinuous than the old (figure V.5).

The building of dams can lead to channel aggradation upstream from the reservoir and channel deepening downstream because of the changes brought about in sediment loads. The overall effect of the creation of a reservoir by the construction of a dam is to lead to a reduction in downstream channel capacity, of about 30–70 per cent.

Equally far-reaching changes in channel form are produced by land-use changes and the introduction of soil conservation measures. Figure V.6 is an idealized representation of how the river basins of Georgia, USA have been modified through human agency between 1700 (the time of European settlement) and the present. Clearing of the land for cultivation (figure V.6(b)) caused massive slope erosion which

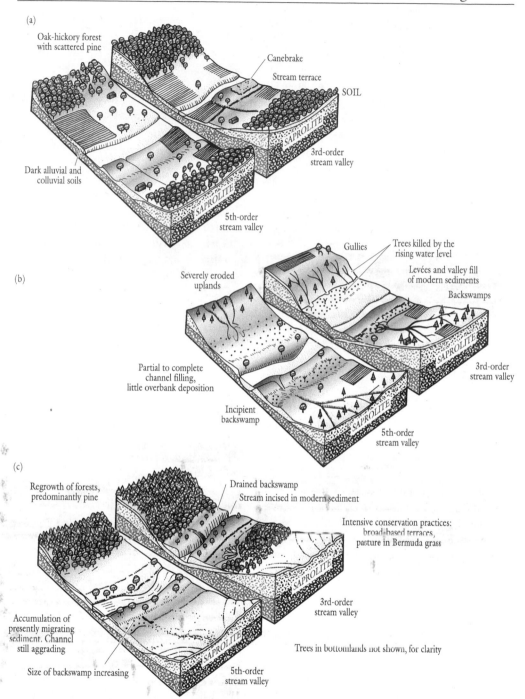

(a)

Oak-hickory forest
with scattered pine

Canebrake

Stream terrace

SOIL

SAPROLITE

3rd-order
stream valley

Dark alluvial and
colluvial soils

SAPROLITE

5th-order
stream valley

(b)

Gullies

Trees killed by the
rising water level

Levées and valley fill
of modern sediments

Backswamps

Severely eroded
uplands

SAPROLITE

3rd-order
stream valley

Partial to complete
channel filling,
little overbank deposition

Incipient
backswamp

SAPROLITE

5th-order
stream valley

(c)

Regrowth of forests,
predominantly pine

Drained backswamp

Stream incised in modern sediment

Intensive conservation practices:
broad-based terraces,
pasture in Bermuda grass

SAPROLITE

3rd-order
stream valley

Accumulation of
presently migrating
sediment. Channel
still aggrading

Trees in bottomlands not shown, for clarity

SAPROLITE

Size of backswamp increasing

5th-order
stream valley

Figure V.6 Changes in the evolution of the fluvial landscapes of the Piedmont of Georgia, USA, in response to land-use change between 1700 and 1970: (a) at the time of European settlement, c.1700; (b) after the clearing and erosive cultivation of uplands; (c) after the checking of erosion and the consequent incision of the headwater stream

Source: After Trimble (1974), p. 117.

resulted in large quantities of sediment being moved into channels and floodplains. Intense erosive land use continued and was particularly strong during the nineteenth century and the first decades of the twentieth century. Thereafter (figure V.6(c)) conservation measures, reservoir construction and a reduction in the intensity of agricultural land use led to further channel changes (Trimble, 1974). Streams ceased to carry such a heavy sediment load and became much less turbid. As a result they cut down into the floodplain sediments of modern alluvium, lowering their beds by as much as 3–4 metres. Another cause of significant changes in channels is the accelerated sedimentation associated with changes in the vegetation growing alongside the channels. In the southern USA the introduction of a bushy tree, the salt cedar, has caused significant

floodplain aggradation. In the case of the Brazos River in Texas, for example, the plants encouraged sedimentation by their damming and ponding effect. They clogged channels by invading sand banks and sand bars, and so increased the area subject to flooding. Between 1941 and 1979 the channel width declined from 157 metres to 67 metres, and the amount of aggradation was as much as 5.5 metres (Blackburn et al., 1983).

Finally, mining can lead to very major changes in channel morphology. The reason for this is that mining often requires the use of large quantities of water and produces large quantities of waste. The waste tends to lead to the aggradation of the channel bed, and if the waste material is coarse then there may be a tendency for a natural meandering pattern to be replaced by a braiding one.

FURTHER READING

Brookes, A., 1988, *Channelized Rivers*. Chichester: Wiley.
An advanced research monograph with broad scope.

5   SALINIZATION

Salinity is a normal and natural feature of soils, especially in dry areas. However, various human activities are increasing its extent and severity.

Salinity in soils has a range of undesirable consequences. For example, as irrigation water is concentrated by evapotranspiration, calcium and magnesium components tend to **precipitate** as carbonates, leaving sodium ions dominant in the soil solution. The sodium ions tend to be absorbed on to **colloidal** clay particles, **deflocculating** them and leaving the soil structureless, almost impermeable to water and unfavourable to root development. Poor soil structure and toxicity lead to the death of vegetation in areas of saline patches. This creates bare ground

which is vulnerable to erosion by wind and water.

Probably the most serious result of salinization is its impact on plant growth. This takes place partly through its effect on soil structure, but more significantly through its effects on **osmotic** pressures and through direct toxicity. When a water solution containing large quantities of dissolved salts comes into contact with a plant cell it causes the cell's protoplasmic lining to shrink. This is due to the osmotic movement of the water, which passes out from the cell towards the more concentrated soil solution. The cell collapses and the plant dies.

This toxicity effect varies with different plants and different salts. Sodium carbonate, by creating highly alkaline soil conditions, may damage plants by a direct caustic effect; high nitrate may promote undesirable vegetative growth in grapes or sugar beets at the expense of sugar content. Boron is injurious to many crop plants at solution concentrations of more than 1 or 2 ppm.

There are a variety of reasons why soil salinity is spreading. The most important of these is the growth in the area of irrigated land, which has increased from about 8 million hectares in 1800 to 250 million hectares in the 1990s (Thomas and Middleton, 1993). The extension of irrigation and the use of a wide range of different techniques for water abstraction and application can lead to a build-up of salt levels in the soil. This happens because water abstraction raises the ground-water level so that it is near enough to the ground surface for water to rise to the surface by capillary action. Evaporation then leaves the salts in the soil. In the case of the semi-arid northern plains of Victoria in Australia, for instance, the water table has been rising at around 1.5 metres per year so that now, in many areas, it is little more than 1 metre below the surface. When groundwater comes within 3 metres of the surface in clay soils – less for silty and sandy soils – capillary forces bring moisture to the surface where evaporation takes place, leaving salts behind.

Second, many irrigation schemes spread large quantities of water over the soil surface. This is especially true for rice cultivation. Such surface water is readily evaporated, so that again salinity levels build up.

Third, the construction of large dams and barrages to control water flow and to give a head of water creates large reservoirs from which further evaporation can take

Plate V.7 'A satanic mockery of snow.' Waterlogged and salinized land in Sind, Pakistan. The white surface is not snow, but salt, a major cause of declining agricultural yields. (A. S. Goudie)

place. The water gets saltier. This salty water is then used for irrigation, with the effects described in the previous paragraph.

Fourth, water seeps laterally from irrigation canals, especially in highly permeable soils, so that further evaporation takes place. Many distribution channels in a gravity irrigation scheme are located on the elevated areas of a floodplain or riverine plain to make maximum use of gravity. The elevated landforms selected are natural levées, river-bordering dunes and terraces, all of which are composed of silt and sand which may be particularly prone to loss by seepage.

In coastal areas salinity problems are created by seawater incursion brought about by overpumping of fresh groundwater from aquifers. If the aquifer is open to penetration from the sea, salty water tends to replace the freshwater that has been extracted. This is a particularly serious problem along the shores of the Persian Gulf where, because of the dry climate, the freshwater can only slowly be replenished by rainfall. However, it can be a problem for any coastal aquifer.

Increases in soil salinity are not restricted to irrigated areas. In some parts of the world salinization has resulted from vegetation clearance (Peck, 1978). The removal of native forest vegetation allows more rainfall to penetrate into deeper soil layers. This causes groundwater levels to rise, creating seepage, sometimes of saline water in low-lying areas. Through this mechanism an estimated 200,000 hectares of land in southern Australia, which at the start of European settlement in the late eighteenth century supported good crops of pasture, is now suitable only for **halophytic** species. Similar problems exist also in North America, notably in Manitoba, Alberta, Montana and North Dakota.

The clearance of the native evergreen forest (predominantly *Eucalyptus* forest) in south-western Australia has led both to

an increase in recharge rates of groundwater, and to an increase in the salinity of streams as salty groundwater seeps out from the ground and into stream flow. Replanting has been shown to reverse the process (Bari and Schofield, 1992).

Salinity can also be increased by saline materials transferred from lake beds that have dried up because of inter-basin water transfers. Around 30–40 million tonnes of salty soils are blown off the Aral Sea every year (see part IV, section 9), for example, and these add to the salt content of soils downwind.

It has been estimated (table V.7) that salt-affected and waterlogged soils account for 50 per cent of the irrigated area in

Table V.7   Salinization of irrigated cropland in selected countries

Country	% of irrigated lands affected by salinization
Algeria	10–15
Australia	15–20
China	15
Colombia	20
Cyprus	25
Egypt	30–40
Greece	7
India	27
Iran	<30
Iraq	50
Israel	13
Jordan	16
Pakistan	<40
Peru	12
Portugal	10–15
Senegal	10–15
Sri Lanka	13
Spain	10–15
Sudan	20
Syria	30–35
USA	20–25

Source: Gleick (1993), table E.5.

Iraq, up to 40 per cent of all Pakistan, 50 per cent in the Euphrates Valley of Syria, 30–40 per cent in Egypt and up to 30 per cent in Iran. In Africa, however, where there are fewer great irrigation schemes, less than 10 per cent of salt-affected soils are so affected because of human action (Thomas and Middleton, 1993). Looking at the problem on a global basis, the calculations of Rozanov et al. (1990) make grim reading. They estimate (p. 210): 'From 1700 to 1984, the global areas of irrigated land increased from 50,000 to 2,200,000 km², while at the same time some 500,000 km² were abandoned as a result of secondary salinization.' They believe that in the last three centuries irrigation has resulted in 1 million sq km of land destroyed, plus 1 million sq km of land with diminished productivity due to salinization. Given the seriousness of the problem, a range of techniques for the eradication, conversion or control of salinity have been developed. These have been reviewed by Rhoades (1990), and include the following:

- provision of adequate subsoil drainage to prevent waterlogging, to keep the water table low enough to reduce the effects of capillary rise and to remove water that is in excess of crop demand;
- leaching of salts by applying water to the soil surface and allowing it to pass downward through the root zone;
- treatment of the soil (with additions of calcium, magnesium, organic matter, etc.) to maintain soil permeability;
- planting of crops which do not need much water;
- planting of crops or crop varieties that will produce satisfactory yields under saline conditions;
- reduction of seepage losses from canals and ditches by lining them (e.g. with concrete);
- reduction in the amounts of water applied by irrigation, by using sprinklers and tricklers;
- storage of heavily salted waste water from fields in evaporation ponds.

FURTHER READING

Worthington, E. B. (ed.), 1977, *Arid Land Irrigation in Developing Countries: Environmental Problems and Effects*. Oxford: Pergamon.
A collection of papers that was among the first and most persuasive considerations of the problem caused by the rapid spread of irrigation schemes.

6   ACCELERATED LANDSLIDES

In 1963, 2,600 people were killed in Italy when a great landslide fell into a reservoir and caused a mass of water to spill over the dam and cascade downstream. Three years later, at Aberfan in South Wales, a massive debris flow killed over 150 people when it destroyed a school and houses as it ran down from a steep coal-waste tip. These are just two of the worst examples of how human actions have created hazardous mass movements on slopes.

Human capacity to change a hillside and to make it more prone to failure has been transformed by engineering development. Excavations are going deeper, buildings and other structures are larger, and many sites which are at best marginally suitable for engineering projects are now being used because of increasing pressure on land. This applies especially to some of the expanding urban areas in the humid parts of low latitudes – Hong Kong, Kuala Lumpur, Rio de Janeiro and many others. Mass movements are very seldom deliberately accelerated by human agency. Most are accidentally caused, the exception possibly being the deliberate triggering of a threatening snow avalanche.

The forces producing slope instability and landsliding can usefully be divided into disturbing factors and resisting properties. Some disturbing factors are natural; others, marked with an asterisk in the following list, are caused by humans.

- *Removal of lateral or underlying support:*
  undercutting by water (for example, river, waves), or glacier ice;
  weathering of weaker strata at the toe of the slope;
  washing out of granular material by seepage erosion;
  *human cuts and excavations, drainage of lakes or reservoirs.
- *Increased disturbing forces:*
  natural accumulations of water, snow, **talus**;
  *pressure caused by human activity (for example, stockpiles of ore, tip-heaps, rubbish dumps, or buildings).
- *Transitory earth stresses:*
  earthquakes;
  *continual passing of heavy traffic.
- *Increased internal pressure:*
  build-up of pore-water pressures (for example, in joints and cracks, especially in the tension crack zone at the rear of the slide).

Factors leading to a decrease in the resisting properties (shear strength) of the materials making up a slope can also be summarized, as follows. Again, those resulting from human activity are marked with an asterisk.

- *Materials:*
  beds which decrease in shear strength if water content increases (clays, shale, mica, schist, talc, serpentine), for example, *when local water table is artificially increased in height by reservoir

construction; or as a result of stress release (vertical and/or horizontal) following slope formation;
low internal cohesion (for example, consolidated clays, sands, porous organic matter).
In bedrock: faults, bedding planes, joints, foliation in schists, cleavage, **brecciated** zones, and pre-existing shears.
- *Weathering changes:*
  weathering reduces effective cohesion, and to a lesser extent the angle of shearing resistance;
  absorption of water leads to changes in the fabric of clays (for example, loss of bonds between particles or the formation of fissures).
- *Pore-water pressure increase:*
  High groundwater table as a result of increased precipitation, or *as a result of human interference (for example, *dam construction) (see under 'Materials' above).

Some mass movements are created by humans piling up waste soil and rock into unstable accumulations that fail spontaneously. The disaster at Aberfan, in South Wales, referred to at the beginning of this section, occurred when a pile of coal waste 180 metres high began to move as an earth flow. The pile had been constructed not only with steep slopes but also upon a spring line.

In the case of the Vaiont Dam disaster in Italy (also referred to at the beginning of this section), heavy rainfall and the presence of young, highly folded sedimentary rocks provided the necessary predisposing conditions for a slip to take place. However, it was the construction of the Vaiont Dam itself which changed the local groundwater conditions sufficiently to affect the stability of a rock mass on the margins of the reservoir. The result was that 240 million cu metres of ground

Table V.8 Examples of methods of controlling mass movements	
Type of movement	*Method of control*
Falls	Flattening the slope Benching the slope Drainage Reinforcement of rock walls by grouting with cement, anchor bolts Covering of wall with steel mesh
Slides and flows	Grading or benching to flatten the slope Drainage of surface water with ditches Sealing surface cracks to prevent infiltration Subsurface drainage Rock or earth buttresses at foot Retaining walls at foot Pilings through the potential slide mass

Source: Dunne and Leopold (1978), table 15.16.

slipped with enormous speed into the reservoir, producing a sharp rise in water level which spilled over the dam, causing flooding and loss of life downstream.

It is evident from what has been said about the predisposing causes of the slope failure *triggered* by the Vaiont Dam that human agency was only able to have such an impact because the natural conditions were broadly favourable to such an outcome.

Although the examples of accelerated mass movements that have been given here are associated with the effects of modern construction projects, more long-established activities, including deforestation and agriculture, are also highly important. For example, Innes (1983) has demonstrated, on the basis of the size of lichens developed on debris-flow deposits in the Scottish Highlands, that most of the flows have developed in the last 250 years. He suggests that intensive burning and grazing may be responsible. Present-day deforestation can generate spectacular mass movements.

Because of the hazards presented by mass movements, a whole series of techniques have been developed to attempt to control them (table V.8).

FURTHER READING

Cooke, R. U. and Doornkamp, J. C., 1993, *Geomorphology in Environmental Management*, 2nd edn. Oxford: Oxford University Press.
This general text contains useful material on slope problems and their control.

Dikau, R., Brunsden, D., Schrott, L. and Ibsen, M-L., 1996, *Landslide Recognition*. Chichester: Wiley.
An edited text, rich in European examples, which describes and classifies the main types of landslides that pose hazards to human activities.

Slope erosion in the Pacific north-west of North America

The mountainous regions of Oregon, Washington, British Columbia and Alaska are areas with steep slopes, high rainfall and active tectonics. They are thus areas of high potential erosion rates. Heavy forest vegetation and the high infiltration capacities of many forest soils protect the slopes; however, the removal of forest in the area, and road-building to take the timber out, have had a series of profound effects. Studies by Swanston and Swanson (1976) have shown a dramatic increase in the occurrence of violent debris avalanches, flows and slides (table V.9). These shallow mass movements leave scars in the form of spoon-shaped depressions from which up to 10,000 cu metres of soil and organic material have moved downslope. They may move as fast as 20 metres per second. Clear-cutting of forest results in an acceleration by two to four times of debris avalanche erosion, while road construction could accelerate debris avalanche erosion by between 25 and 340 times the rate under undisturbed forest!

Table V.9   Debris-avalanche erosion in forest, clear-cut and roaded areas

Site	Period of records (years)	Area (sq km)	No. of slides	Debris-avalanche erosion (cu metres/ sq km/yr)	Rate of debris-avalanche erosion relative to forested areas
Stequaleho Creek, Olympic Peninsula					
Forest	84	19.3	25	71.8	× 1.0
Clear-cut	6	4.4	0	0	0
Road	6	0.7	83	11,825	×165
Total	–	24.4	108		
Alder Creek, western Cascade Range, Oregon					
Forest	25	12.3	7	45.3	× 1.0
Clear-cut	15	4.5	18	117.1	× 2.6
Road	15	0.6	75	15,565	×344
Total	–	17.4	100	–	–
Selected drainages, Coast Mountains, south-west British Columbia					
Forest	32	246.1	29	11.2	× 1.0
Clear-cut	32	26.4	18	24.5	× 2.2
Road	32	4.2	11	282.5	× 25.2
Total	–	276.7	58	–	–
H. Andrews Experimental Forest, western Cascade Range, Oregon					
Forest	25	49.8	31	35.9	× 1.0
Clear-cut	25	12.4	30	132.2	× 3.7
Road	25	2.0	69	1,772	× 49
Total	–	64.2	130	–	–

7 GROUND SUBSIDENCE

Like many of the environmental issues discussed in this book, ground subsidence can be an entirely natural phenomenon. For example, climatic change may cause permanently frozen subsoil (permafrost) to decay in tundra areas: this will produce swampy depressions called thermokarst. Likewise, in limestone areas true **karstic** phenomena, such as swallow holes, may develop when the ground surface collapses into a subterranean cavity produced by the solution of limestone over a long period.

Nevertheless, humans are now causing ground subsidence to occur at an accelerating rate, and with dramatic consequences, in certain sensitive areas. The main mechanisms are:

- the transfer and removal of subterranean fluids such as oil, gas and water;
- the removal of solids, either through underground mining (e.g. coal and other minerals) or in solution (e.g. salt);
- the disruption of permafrost;
- the compaction or reduction of sediments (especially organic-rich ones) by irrigation and land drainage;
- the construction of reservoirs.

Ground subsidence is often a relatively gentle progress, but sometimes it can be sudden and catastrophic. This is particularly the case in areas where the bedrock is limestone and where overpumping has greatly drawn down the level of the water table. A sensitive area of this kind is the Far West Rand of the Transvaal in South Africa, where gold mining has required that the local water table be lowered by more than 300 metres. The fall of the water table has caused clay-rich materials filling the roofs of large underground caves to dry out, shrink and collapse. This in turn has caused large depressions to develop at the ground surface. In densely populated urban areas this is a considerable hazard. In another limestone area, Alabama in the southern USA, groundwater pumping has caused over 4,000 **sink-holes** or related features to form since 1900. Fewer than 50 natural sink-holes developed in that area over the same period.

More gentle, but in geological terms still very rapid, has been ground subsidence caused by oil abstraction. The classic case is the Los Angeles area, where over 9 metres of subsidence occurred as a result of the development of the Wilmington oilfield between 1928 and 1971. Consider that 9 metres is more than the average height of a two-storied house! Even more widespread is the subsidence caused by groundwater abstraction for industrial, agricultural and domestic purposes. In Mexico City subsidence of more than 7.5 metres has occurred, while in the Central Valley of California the figure exceeds 8.5 metres. In Tokyo, Japan, subsidence has brought some areas below sea level. In 1960 only 35 sq km of the Tokyo lowland was below sea level. By 1974 this figure had almost doubled, exposing a total of 1.5 million people to major flood hazard. Bangkok is suffering from a similar problem.

Perhaps the most familiar example of ground subsidence caused by humans is that resulting from mining. It causes damage to houses, roads and other structures as well as disrupting surface drainage and causing flooding.

In permafrost areas ground subsidence is associated with thermokarst development. Thermokarst is the irregular, hummocky terrain produced by the melting of ground ice, permafrost. The development of thermokarst is due primarily to the disruption of the thermal equilibrium of the permafrost and an increase in the depth of the active layer (the layer subjected to annual thawing). Consider an undisturbed

with an active layer of 45 cm.
⟩ that the soil beneath 45 cm
ırated permafrost and upon
lds (on a volume basis) 50 per
cent water and 50 per cent saturated soil.
If the top 15 cm were removed, the equilibrium thickness of the active layer, under the bare ground conditions, might increase to 60 cm. As only 30 cm of the original active layer remains, 60 cm of the permafrost must thaw before the active layer can thicken to 60 cm, since 30 cm of **supernatant** water will be released. Thus, the surface subsides 30 cm because of thermal melting associated with the degrading permafrost, to produce an overall depression of 45 cm.

Thus the key factors involved in thermokarst subsidence are the state of the active layer and its thermal relationships. When surface vegetation is cleared for purposes of agriculture or construction, for example, the depth of thaw will tend to increase as the ground will no longer be insulated from the effects of direct sunlight. The movement of tracked vehicles has been particularly harmful to surface vegetation and deep channels may soon result from permafrost degradation where these have been used. Similar effects may be produced by siting heated buildings on permafrost, and by laying oil, sewer and water pipes in or on the active layer.

Some subsidence is created by a process called **hydrocompaction**. This occurs because moisture-deficient, unconsolidated, low-density sediments tend to have sufficient dry strength to support considerable effective stresses without compacting.

However, when such sediments, which may include alluvial fan materials or loess, are thoroughly wetted for the first time (for example, by percolating irrigation water), the inter-granular strength of the deposits is diminished. Rapid compaction results, and subsidence of the ground surface follows. Unequal subsidence can create problems for irrigation schemes.

Land drainage can promote subsidence of a different type, notably in soils rich in organic matter. The lowering of the water table makes peat susceptible to oxidation and deflation (being blown away by the wind in dust storms), so that its volume decreases. We discuss this in the context of the English Fenlands in part IV, section 5.

A further type of subsidence, sometimes associated with earthquake activity, results from the effects on the earth's crust of large masses of water impounded behind dams. Seismic effects can be generated in areas with susceptible fault systems. This may account for earthquakes recorded at Koyna (India) and elsewhere. The process whereby a mass of water causes crustal depression is called **hydro-isostasy**.

It is clear from this discussion that ground subsidence is a diverse but important facet of the geomorphological impact of human activity. The damage caused on a worldwide basis can be measured in billions of dollars each year. We have mentioned some of the forms such damage takes in this section. They include broken dams, cracked buildings, offset roads and railways, fractured well casings, deformed canals and impeded drainage, among many others.

FURTHER READING

Johnson, A. T. (ed.), 1991, *Land Subsidence*. IAHS Publication no. 200.
A large collection of research-level papers.

Waltham, A. C., 1991, *Land Subsidence*. Glasgow: Blackie.
A lower-level introductory study, which is particularly strong on the effects of mining on subsidence.

8   WASTE DISPOSAL

Waste can be loosely defined as 'all un-used, unwanted and discarded materials including solids, liquids, and gases' (Costa and Baker, 1981, p. 397). Alternatively, it can be defined as 'something for which we have no further use and which we wish to get rid of'. However it is defined, there is no doubt that waste is generated in large quantities by humans, that the amount of waste generated develops as societies become greater consumers of materials, and that the control and disposal of waste is a

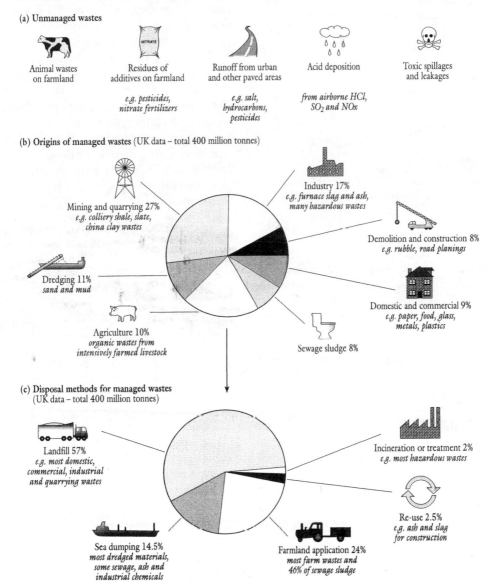

Figure V.7   The main sources of, and disposal strategies for, wastes that can pollute land and water, early 1990s
Source: After Woodcock (1994), fig. 16.1.

Table V.10   Wastes arising in England and Wales, late 1980s

Waste type	Quantity (mt/yr)
Liquid industrial effluent	2,000
Agricultural	250
Mining and quarrying	130
Industrial	50
Hazardous	3.9
Special	1.5
Domestic and trade	28
Sewage sludge	24
Power station ash	14
Blast furnace slag	6
Building	3
Total	2,505

Source: House of Commons Environment Committee, *Second Report, Toxic Waste* (1988/9).

major environmental issue. Furthermore, the disposal of waste can substantially modify surface conditions and produce an array of environmental impacts.

Wastes can be divided into those that are unmanaged and those that are managed (figure V.7(a) and (b)).

Table V.10 shows the amounts of different categories of waste produced in England and Wales. In terms of sheer weight, liquid industrial effluent is the largest component, but the production of effluent from the agricultural sector is also important. Significant amounts of primarily solid waste arise from mining and quarrying, industrial sources, the domestic sector, sewage sludge, power station ash, blast furnace slag, and the building and construction industries. In the USA an average city with 250,000 inhabitants has to collect, transport and dispose of 450 tonnes of refuse every day. In general about 2–3 kg of municipal waste and 3–4 kg of industrial waste are produced in the USA per person

per day. In the UK, about 137 million tonnes of 'controlled waste' (waste which is either incinerated or disposed of to a landfill) are produced every year. Landfill takes 90–95 per cent of the controlled waste.

In this section we are mainly concerned with solid waste, a category which includes materials from a wide range of sources (table V.11). There are a number of disposal options for solid waste (figure V.7 (c)). As we have already noted, the most important of these in the UK is so-called 'sanitary landfill' (table V.12). The relative importance of different methods varies from country to country (table V.13). For example, whereas most municipal solid waste in the UK and Australia goes to landfills, in Switzerland nearly half is incinerated and in Japan about two-thirds is incinerated.

The content of waste is also highly variable. Some types contain contaminants that can pose a series of hazards to health and property (table V.14). For example, industrial waste may contain dangerous heavy metals, building waste may contain asbestos, and household food waste may generate potentially explosive methane gas. If sites are not carefully controlled, waste draining from the site (**leachate**) may be heavily polluted. Other problems are posed by the fact that landfill may gradually compact through time.

Landfilling is a cheap means of disposal. In countries where there are many old quarries and gravel pits, it may be a convenient way not only to dispose of waste but also to reclaim such land for other uses. Such sites are not always available, however, in close proximity to sources. Also, if they are badly managed they can produce environmental problems of the types just discussed.

There may be advantages in reducing the amount of landfill capacity that is required. A range of methods is available:

Table V.11   Refuse materials (solid waste)

Type	Composition	Sources
Garbage	Wastes from preparation, cooking and serving of food, market wastes, wastes from handling, storage and sale of produce	Households, restaurants, institutions, stores, markets
Rubbish	Combustible paper, cartons, boxes, barrels, wood, shavings, tree branches, yard trimmings, wood furniture, bedding	Same as garbage
Ashes	Residue from fires used for cooking and heating and from on-site incineration	Same as garbage
Street refuse	Sweepings, dirt, leaves, catch-basin dirt, contents of litter receptacles	Streets, sidewalks, alleys, vacant lots
Dead animals	Cats, dogs, horses, cows	Same as street refuse
Abandoned vehicles	Unwanted cars and trucks left on public property	Same as street refuse
Industrial wastes	Food-processing wastes, boiler-house cinders, lumber scraps, metal scraps, shavings	Factories, power plants
Demolition wastes	Lumber, pipes, brick, masonry, and other construction materials from razed buildings and other structures	Demolition sites to be used for new buildings, renewal projects, expressways
Construction wastes	Scrap lumber, pipe, other construction materials	New construction, remodelling
Special wastes	Hazardous solids and liquids: explosives, pathological wastes, radioactive materials	Households, hotels, hospitals, institutions, stores, industry
Sewage treatment residue	Solids from coarse screening and from grit chambers; septic-tank sludge	Sewage treatment plants, septic tanks

Source: Costa and Baker (1981), table 13–1. Data from Institute for Solid Wastes of the American Public Works Association and Bureau of Solid Waste Management, 1970.

Table V.12 Methods of solid waste disposal	
Method	*Description*
Open dumps	Practices vary from indiscriminate piles to periodic levelling and compacting
	Little effort taken to prevent rodents, flies, odours, and other health hazards
	Often located with little planning where land was available
Sanitary landfills	Consists of alternating layers of compacted refuse and soil. Each day refuse is deposited, compacted, and covered with soil
	Daily operation and a final cover of at least 70 cm of compacted soil prevents many health problems
	Requires planning for economical operation and for supplies of topsoil for cover. Operations vary depending on topography and supplies
Incineration	Reduces combustible waste by burning at 1,700°F to an inert residue. Ash and noncombustibles dumped or placed in landfills
	Air pollution is a problem with poor management
	Increasing in use and often combined with a sanitary landfill and salvage operation
On-site disposal	Small-scale incinerators and garbage disposals
	Incinerators are expensive and require considerable maintenance
	Garbage disposals are increasing rapidly in use, with waste transferred directly to the sanitary-sewer system
Swine feeding	A decreasingly used method which involves collection of garbage for swine food (pig swill)
Composting	Biochemical decomposition of organic materials to a humus-like material usually carried out in mechanical digesters
	Increasingly used method with a useful end product which is often sold

Source: Costa and Baker (1981), table 13–3; from Schneider (1970).

Table V.13   Selected solid waste material statistics for various countries

Country	Annual per capita production (kg)	% disposed by landfill	% disposed by incineration
Australia	681	98	2
Austria	216	57	19
Canada	642	94	6
Denmark	420	64	32
France	289	33	32
Germany (W.)	447	83	9
Italy	246	38	20
Japan	342	28	67
Netherlands	502	66	19
Sweden	300	52	38
Switzerland	336	13	49
UK	332	80	6
USA	744	n/a	n/a

Source: UNEP (1990).

Table V.14   Some commonly encountered contaminants, the sites on which they are likely to occur and the principal hazards they produce

Type of contaminant	Likely to occur	Principal hazards
'Toxic' metals e.g. cadmium, lead, arsenic, mercury	Metal mines, iron and steel works, foundries, smelters Electroplating, anodizing and galvanizing works	Harmful to health of humans or animals if ingested directly or indirectly. May restrict or prevent the growth of plants
Other metals e.g. copper, nickel, zinc	Engineering works, e.g. shipbuilding. Scrap yards and shipbreaking sites	
Combustible substances, e.g. coal and coke dust	Gasworks, power stations, railway land	Underground fires
Flammable gases e.g. methane	Landfill sites, filled dock basins	Explosions within or beneath buildings
'Aggressive' substances e.g. sulphates, chlorides, acids	'Made ground' including slags from blast furnaces	Chemical attack on building materials e.g. concrete foundations
Oily and tarry substances, phenols	Chemical works, refineries, by-products plants, tar distilleries	Contamination of water supplies by deterioration of service mains
Asbestos	Industrial buildings. Waste disposal sites	Dangerous if inhaled

Source: Attewell (1993), table 4.1.

Plate V.8   Landfill is one way of disposing of waste, but the choice of sites to use can be a problem. This site is filling in old gravel pits near Didcot, central England.

- *Incineration* can greatly reduce the volume of waste. However, incinerators are expensive to construct and may create pollutant emissions to the air. Concerns have been expressed, for example, about dioxin emissions. On the positive side, incinerators can produce usable energy.
- *Compaction* can also reduce waste volume. Powerful hydraulic rams can be used to compress waste.
- *Shredding and baling* can also reduce waste volume.

However, it may be more desirable to reduce the amount of waste produced in the first place. This can be achieved by:

- substituting durable goods for disposable ones;
- composting garden waste;
- generating less waste;
- reusing materials and extending their lives (e.g. by using rechargeable batteries and refillable bottles);
- recycling paper, glass, etc.;
- recovering materials from waste (e.g. magnetic separation of ferrous metals).

FURTHER READING

Douglas, T., 1992, Patterns of land, water and air pollution by wastes. In M. Newson (ed.), *Managing the Human Impact on the Natural Environment: Patterns and Processes*, 150–71. London: Belhaven Press.
A very useful review chapter in an introductory textbook.

9 STONE DECAY IN URBAN BUILDINGS

The natural materials we use for building are just as prone to weathering and alteration as are natural rock outcrops. Similarly, manufactured building materials such as bricks, concrete and plastics also decay and change once in contact with the atmosphere. Usually, such decay processes are of no real concern as they act very slowly and produce only slight changes to the appearance of buildings, monuments and engineering structures and do not affect their strength, safety or economic life-span. However, where decay processes become accelerated and altered, usually as a result of air pollution, they can obliterate priceless carvings, produce unsightly decay features, and lead to structural weakness. Many buildings and monuments are at risk, from the historic basilica of St Mark's in Venice to Lincoln Cathedral in England, the Parthenon in Athens and the Merchants Exchange Building in Philadelphia. In many cities whole groups of buildings and monuments are under attack. Examples are the historic university town of Oxford in England and the beautiful city of Prague in the Czech Republic.

Buildings in the urban environment are particularly vulnerable to decay because of the following factors.

- urban **microclimatic** changes, such as warming and increased local rainfall or humidity;
- air pollution, such as increased concentrations of sulphur dioxide and nitrogen oxides;
- increased urban traffic levels, which contribute to air pollution, lead to application of de-icing salts in winter in many temperate-zone cities, and cause vibrations affecting roadside buildings;

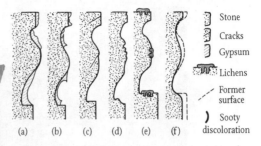

Figure V.8 Some common forms of building stone decay: (a) sooty and gypsum encrustations on sheltered parts of a building; (b) blistering and exfoliation of gypsum crusts from (a); (c) cracking; (d) pitting, blistering and exfoliation of porous stone which has been affected by salts; (e) lichen growths on stone with biological weathering underneath; (f) surface lowering and roughening by acid rain

- increased human contact with buildings, leading to abrasion, graffiti, etc.

Cities on coasts and within the arid zone suffer particularly from highly corrosive environments, because of high concentrations of salt in the atmosphere and groundwater.

These environmental conditions in urban areas produce the following effects on building and monument surfaces (see figure V.8):

- gypsum crusts, produced by direct chemical reaction of sulphur dioxide with calcium carbonate-rich stone;
- soiling of building materials by sooty particulates commonly produced by coal and oil combustion and diesel engines;
- accelerated lowering of surfaces, produced by acidified rainfall hitting calcium carbonate-rich stone;

Plate V.9   Decaying buildings in Venice. The sheltered portion of the column shows the development of a black crust which contrasts sharply with the light colour of the portion of the column that is washed by rain. (Dr B. Smith)

- exfoliation and blistering, produced by salt crystallization and hydration in porous materials;
- cracking, produced by vibration and other stresses;

- pitting and surface growths, produced by organic colonization, especially by micro-organisms and lichens, possibly encouraged by nitrogen oxides acting as fertilizers.

Vast changes in pollution and environment have occurred in many cities over recent years. These changes have had consequent impacts on the weathering and decay of buildings and monuments. Thus, a rapid increase in urban traffic and photochemical pollution in Athens seems to have been reflected in increasing stone damage on the many ancient marble monuments in the city. In other cities, such as those in Britain and the USA, legislation to combat air pollution has resulted in sharp decreases in sulphur dioxide and smoke pollution over the past 40 years, but not in nitrogen oxides. Measurements from St Paul's Cathedral in London, however, indicate that the rate of decay in building stone has not yet declined. Indeed, stone decay may worsen in some places, as nitrogen oxides act as a fertilizer for organic growths such as bacteria and lichens, which are important agents of stone decay.

How serious a problem is urban stone decay, and what can we do about it? In terms of cost, stone decay can be a serious problem for the owners of buildings, especially when it has turned into a long-term problem. In Oxford, England, for example, by the middle of the twentieth century 200 years or so of air pollution from domestic and industrial coal burning had produced intense damage to buildings constructed from the rather vulnerable Headington Freestone (a local limestone which weathers badly in polluted atmospheres). Restoration work costing over £2.4 million was carried out in the 1950s and 1960s. More recently, traffic and other sources of air pollution are damaging these restored buildings, as well as new ones, and soiling paintwork within the city centre. Stone decay is particularly serious when it affects monuments of great cultural and spiritual significance, especially those which attract large numbers of tourists and their associated income. Decay can also be hazardous, as when it affects bridges, or causes bits of stone to fall from high towers. In most cities, however, building stone decay is just one symptom of increasing urban pollution and environmental despoliation. The impacts of air pollution on human health and urban ecology in cities are also of great concern (as discussed in part II, section 8 and part III, section 6).

Strategies for combating urban building stone decay include:

- removing the causes of accelerated decay, by reducing air pollution, stopping the application of de-icing salts to roads, etc.;
- removing valuable and vulnerable sculptures and carvings from the aggressive urban environment, putting them in controlled, museum environments and replacing them with copies;
- cleaning and repairing soiled and damaged buildings;
- preventing future decay by applying protective treatments on new or newly cleaned and repaired stone.

FURTHER READING

Winkler, E. M., 1975, *Stone: Properties, Durability in Man's Environment*. Vienna: Springer-Verlag.
Contains much information about many aspects of stone weathering.

Cooke, R. U. and Gibbs, G., 1994, *Crumbling Heritage: Studies of Stone Weathering in Polluted Atmospheres*. Report for National Power plc.
A useful summary of the recent worries over stone decay in Britain and results from research aimed at elucidating the problem.

Venice's decaying treasures

Venice in Italy contains many important buildings and monuments which
form a key part of the European cultural heritage and which are under
threat from decay accelerated by air
pollution and rising sea levels. There
are also over 2,000 pieces of 'outdoor
art', mainly stone carvings and sculptures, within the city. Studies of old
photographs have revealed that most
decay has occurred since the Second
World War. The cause seems to be
the high sulphur dioxide levels resulting from rapid post-war industrialization of the surrounding area (Del
Monte and Vittori, 1985). Since
1973 laws have banned the use of
oil within the city itself, replacing it
with methane. However, pollution

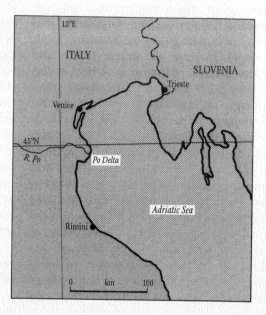

still drifts in from elsewhere and much serious decay has already occurred. Smoke
and sulphur dioxide react with marble, limestone and calcareous sandstones to
produce the blackened, gypsum crusts which now coat many famous buildings in
Venice. These crusts are not only unsightly, they are also damaging the underlying stone. A local relative sea-level rise has been a problem over the past century
in Venice. This has been caused by a combination of natural subsidence and
extraction of groundwater (which has now ceased). As well as creating flooding,
such higher sea levels have had a less visible impact on Venice's environment,
through encouraging the penetration of water and salts into vulnerable building
materials. The humid environment also encourages the transformation of calcium
carbonate into gypsum in the presence of sulphur dioxide.

Major research is currently under way into stone decay in Venice, coupled with
many schemes to restore damaged buildings and 'outdoor art'. Estimates of the
costs of restoration suggest that all the sculptures and carvings in Venice could
be restored at a cost of some $US9.5 million. Grime accounts for 15 per cent
of the damage requiring conservation, corrosion or decay accounts for 35 per
cent, and structural problems for the remaining 50 per cent (Carrera, 1993).

Considerable care has to be taken in attempting to clean and restore valuable
objects and buildings. It is essential first to diagnose the major causes of decay
correctly. Only then can the most appropriate solutions be proposed. The Church
of Santa Maria dei Miracoli, for example, has marble decoration slabs which are
badly damaged. Surveys revealed that salts from rising damp were the major cause
of decay, and techniques to remove the salts were developed before any restoration began.

10 CONCLUSION

In this part of the book we have considered the impact that humans have had on the soil, on geomorphic processes and on landforms. We have drawn attention to the unintended acceleration of rates of soil erosion by water caused by a range of human actions, including deforestation, the use of fire and urban growth. Accelerated soil erosion threatens soil fertility and agricultural productivity. It also has other 'off-farm impacts' which include a lowering of water quality. While this has often been seen as a particular problem in developing countries, where it has been identified as one facet of desertification and land degradation, we have also shown that it is a problem demanding attention in the British context. Likewise, accelerated soil erosion by wind has been a major problem not only in the Sahel zone of Africa and China, but also in the technologically advanced farmlands of the USA and the lowlands of Britain. There are, however, a wide range of soil conservation measures that can be adopted to counter both water and wind erosion caused by land-use and land-cover changes.

Land-use and land-cover changes are also responsible for many other geomorphological changes. They affect the form of river channels and the nature of mass movements (including landslides) on slopes. As is the case with soil erosion, numerous methods are available to try to stabilize slopes and so reduce the hazards posed by slope failures.

Soil erosion and accelerated landslides are not the only serious ways in which the Earth's surface materials are transformed. In particular, the spread of irrigation and the removal of trees can lead to one of the most pernicious forms of soil transformation or metapedogenesis: accelerated salinization. This is a major problem for agricultural production, especially in the drier parts of the world. Again, a range of techniques for the eradication, conversion or control of salinity have been developed.

Another form of accelerated geomorphological change that we have identified for a range of environments from the tundra regions to the world's great deserts is ground subsidence. However, in many parts of the world it is not so much the subsidence of the ground that is the problem, but where to put the ever-increasing quantities of waste which we produce. Landfill is one solution, but there are other options including incineration, compaction, shredding and baling. A more fundamental solution is to reduce the amount of waste produced in the first place.

Finally, we draw attention to the fact that humans alter the weathering environment, particularly by subjecting rocks and other building materials to corrosive air. Stone decay, whether in Venice or Oxford, Prague or York, Athens or Agra, is a serious threat to our cultural heritage. Even though the process is slow compared with the accelerated soil erosion mentioned earlier in this part, it can have a serious impact on buildings and monuments.

Overall, the human impacts on the land surface discussed in this section are a rather mixed bag, often spatially limited in extent and often inadvertently caused. They are, nevertheless, serious and show linkages with human impacts on the biosphere and atmosphere. Many technological solutions have been developed to deal with these problems. Nevertheless, as several of our case studies have shown, the success of such schemes depends on the willingness and ability of the people involved, at all levels, to implement and maintain them.

KEY TERMS AND CONCEPTS

accelerated landslides

accelerated soil erosion

dune reactivation

forest soils

hydrocompaction

landfill

mass movements

permafrost

salinity

sand control

seawater incursion

soil conservaton

subsidence

thermokarst

waste

POINTS FOR REVIEW

Why should we be concerned about soil erosion?

How would you seek to control rates of soil erosion by (a) wind and (b) water?

What are the main ways in which humans unintentionally cause river channel characteristics to change?

Why is salinization such an important issue in the world's drylands?

What are the main geomorphological hazards that are being accelerated by human activities?

How, in your own life, could you reduce the need for waste to be disposed of as landfill?

Is there are evidence in your own home area that buildings are suffering from severe weathering? Why might this be?

Oceans, Seas and Coasts

1 INTRODUCTION

Today, almost 3 billion people (about 60 per cent of the world's population) live near coasts, often in large cities. Furthermore, coastal populations are rapidly increasing. In the USA, for example, population density is growing faster in coastal states than inland ones. Nearly half of all building in the USA between 1970 and 1989 occurred in coastal regions, which account for only 11 per cent of the country's total land area. Similar trends are found in many other countries.

Human activity is contributing to a range of local and regional environmental problems in coastal areas. The main environmental impacts along the world's coastline involve disruption to coastal sedimentation pathways through erosion and accelerated deposition; increased flood hazard, through sea-level rise and encouragement of local subsidence; disruption of coastal ecology, through reclamation of land and changing land uses; and coastal pollution. Historically, attempts to manage the coastal ecosystem have involved trying to make the coast more stable and fixed. These have made many environmental problems worse. In recent years an extra dimension has been added to concerns over coastal environmental problems with the threat of accelerated sea-level rise in the future as a result of global warming.

The world's oceans and seas cover over 70 per cent of the Earth's surface and play a vital role in the biosphere. These vast bodies of water are also being affected by a range of human impacts. Pollution is the major worry. Some pollutants come from ships and oil platforms, but most are from onshore sources, reaching the sea via the atmosphere, rivers or coastal outfalls. Fishing and harvesting of marine resources also have adverse consequences for the marine environment, leading to more pollution and also damaging ecosystems. About 53 million tonnes of marine fish are caught

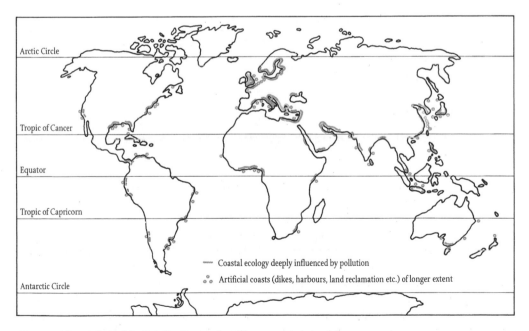

Figure VI.1 World distribution of major coastal problems
Source: After Kelletat (1989).

worldwide for human consumption every year, with an additional 22 million tonnes collected for processing into fish meal, oil, etc. (Tolba and El-Kholy, 1992). At present, the most severe problems are restricted to enclosed seas, such as the Mediterranean and Baltic Seas, surrounded by dense populations. However, the ever-increasing load of pollution entering the world's oceans is likely to cause wider problems in the future.

Coastal areas particularly vulnerable to environmental problems include estuaries, deltas and other low-lying coasts, especially in highly populated areas. Figure VI.1 shows the global distribution of such areas. Parts of the Mediterranean, Baltic, North Sea and Bangladesh coasts are particularly vulnerable to a whole range of problems. Natural and human-induced processes combine to create coastal problems. In most cases these do not occur in isolation, but rather interact to produce a complex web of stresses on the environment. Thus, salt-marsh erosion may be exacerbated by pollution, which interferes with the plant–sediment relationship vital to marsh development; where such erosion occurs, it may encourage flooding. Stresses on one part of the coastline may affect other parts. For example, deforestation can cause mangrove swamp erosion, which in turn leads to downdrift degradation of coral reefs, as they become choked by the extra sediment load. The destruction of coral reefs in turn encourages storm damage to the coastal zone behind the reefs that once sheltered it.

Further Reading

GESAMP, 1990, *The State of the Marine Environment*. Oxford: Blackwell Scientific. An authoritative global survey of marine pollution.

Bird, E. C. F., 1985, *Coastline Changes*. Chichester: Wiley. A country-by-country survey of the erosional state of the world's coastline.

Viles, H. A. and Spencer, T., 1995, *Coastal Problems*. London: Edward Arnold. A textbook which surveys, with many examples, the range of ways in which humans interact with the natural coastal processes.

2   Sea-level Rise

'Sea level' is perhaps a misleading term, for the relative positions of land and oceans are continually changing on a variety of time-scales. These fluctuations range from daily tidal cycles to vast changes in ocean volume related to glacial and interglacial cycles over periods of thousands and millions of years. However, a practical definition of mean sea level is the long-term average (usually over 19 years at least) of high and low tide levels at a particular place. This level is affected by changes in the volume or mass of water (**eustatic** or global changes) and movements of land (tectonic and **isostatic** changes), or a combination of the two. Over the past 18,000 years, since the peak of the last Ice Age, the rising volume of ocean water as the ice on land melted, coupled with complex isostatic changes, has produced a generally increasing mean sea level over the world.

Individual areas, however, have experienced very different sea-level histories (Clark et al., 1978). Over the past 1,000 or so years, sea level has risen (according to a range of evidence) at 0.1–0.2 mm per year. During the last 100 years, sea-level rise has accelerated to 1.0–2.5 mm per year, according to many estimates. This acceleration is mainly due to climatic factors, such as the thermal expansion of ocean waters and the melting of ice on land.

It is predicted that over the next 50 to 100 years global warming will lead to a further acceleration of the rate of sea-level rise through a combination of two effects:

- increased volume as ocean water warms up (called the steric effect);
- addition of water to the oceans from the melting of small glaciers and large ice sheets.

The direct impact of human actions on sea level may also provide an additional acceleration. These actions, and their consequences, include the following:

- extraction of oil and groundwater may encourage coastal subsidence;
- deforestation may encourage increased freshwater runoff to oceans;
- groundwater extraction for irrigation, and damming of rivers to produce reservoirs, may encourage evaporation of this water, which will eventually return to the oceans (Sahagian et al., 1994);
- wetland drainage reduces the water holding capacity of wetland soils, and therefore adds more water to the oceans.

It is very difficult to predict how such influences might combine to affect sea-level rise in the future. The behaviour of some of the compartments of the system (e.g. ice sheets) is not well understood, and the magnitude of global warming in

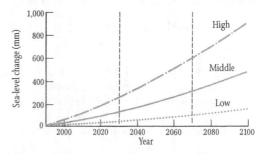

Figure VI.2   Best estimate high, middle and low projections of sea-level rise, to the year 2100, under the IPCC emissions scenario IS92a
Source: After Wigley and Raper (1992), fig. 4.

the next few decades is itself the subject of much debate. However, the most recent estimates suggest an average sea level rise of about 5 mm per year over the next century, within a range of uncertainty of 2–9 mm per year. This will produce a total increase of about 50 cm by 2100, as shown in the middle curve in figure VI.2, and means that sea level will rise two to five times faster than over the last 100 years (Watson et al., 1996). This rate, although high, is much lower than some earlier estimates which predicted widespread drowning of many coastal areas. Whatever its precise magnitude, future sea-level rise, in association with a whole host of smaller human-induced and natural disturbances, is likely to provide a complex series of effects on the coastal environment.

Sea-level rise and the Essex coast marshes, England

Much of the eastern and southern coastline of England is already undergoing relative sea-level rise. Isostatic readjustments to the removal of the ice cap over northern Britain at the end of the last Ice Age (some 11,000 years ago) are causing the north of Britain to rise, forcing the southern part down as a consequence. The Essex coast has been experiencing relative sea-level rise of 4–5 mm per year over the past few decades as a result of such a process. The coastline of Essex is dominated by low-lying estuarine and open coast marshes which play a valuable role in coastal protection, acting as baffles to wave energy and protecting the sea walls on the landward side. Recent sea-level rise, coupled with local human activities, has led to erosion of many of the marshes here.

The Blackwater estuary (figure VI.3) provides a good example of the problems faced by the Essex coast. It has 680 hectares of salt marsh and 2,640 hectares of intertidal mudflats around it, and is backed by agricultural land. The Bradwell nuclear power station is located on its margins. Much of the coastal marshland around the estuary has been reclaimed over the centuries, to increase the area of farmland. Flood embankments now line 95 per cent of the estuary. These

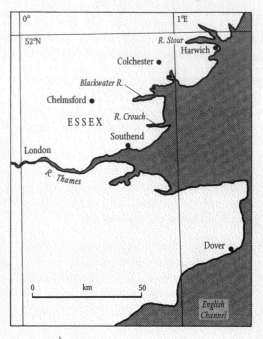

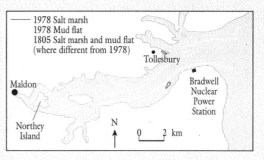

Figure VI.3 The Blackwater estuary, Essex, and its associated marshes and mudflats
Source: After Pethick (1993), figs 1, 2.

limit the ways in which the marshes can react to sea-level rise. Studies show that between 1973 and 1988 23 per cent of the total salt marsh area around the estuary was lost to erosion (Pethick, 1993). Over the past 150 years, sea-level rise here has been accompanied by an increase in width of the main estuary channel and a decrease in its depth. John Pethick, an expert on this particular area of the coast, thinks that accelerated sea-level rise in the future will lead to erosion of marshes in the outer estuary and their replacement by sand and gravel habitats. Further inland, marshes will become less brackish, as salt water penetrates further

Plate VI.1   The salt marshes at Tollesbury, Essex. Marshes such as these will be modified by any future sea-level rise. Marshes nearby are the site of a scheme to stimulate marsh development. (H. A. Viles)

up the estuary. The extensive coastal defences behind the Blackwater marshes mean that, unless humans intervene, sea-level rise will eventually lead to the destruction of these marshes.

As the marshes help to protect the sea walls from coastal erosion, there have been strenuous efforts to help 'save' the marshes. A pilot scheme on Northey Island (whose location is shown in figure VI.3), for example, has pioneered the use of 'set-back' techniques to stimulate new marsh growth by removing an old, broken-down sea wall and allowing the sea to 'reclaim' the land behind it. At Tollesbury, on the northern shore of the estuary, similar experiments are taking place. Here 21 hectares of arable land, bought by English Nature specially for the purpose, is being flooded in a policy of 'managed retreat'. There is a danger that such piecemeal schemes will make the problems worse for the rest of the estuary unless they are carefully managed. John Pethick suggests that a much bigger, estuary-wide project is necessary to manage these vital coastal wetlands in the face of future sea-level rise. Such a scheme would involve allowing the outer estuarine channel to widen, coupled with a general retreat of flood embankments.

Further reading

Pethick, J. S., 1993, Shoreline adjustments and coastal management: physical and biological processes under accelerated sea level rise. *Geographical Journal* 159, 162–8.

3   COASTAL EROSION

Coastal erosion is a natural process, powered by wave energy and vital to the maintenance of a dynamic coastline. Human activity, however, has increasingly been responsible for accelerating coastal erosion. Increasing human settlement near the coast and use of the coastal zone have also created a serious environmental issue which requires sensitive, long-term management.

Recent surveys have produced some stark statistics. For example, net erosion has occurred on over 70 per cent of the world's sandy coastline over the past few decades. However, such erosion does not only affect sandy coastlines. There have been spectacularly high losses of land on the Niger delta. Here, 487 hectares of coastal plain were lost as a result of subsidence caused by oil and gas extraction from the delta, and mangrove deforestation. Where high rates of erosion, however localized they may be (see table VI.1 for Britain), coincide with dense human settlement and intensive coastal use, serious problems result. Along the mid-Atlantic coast of the USA, for example,

barrier islands have retreated at about 1.5 metres per year as sediment from the ocean side is eroded and washed over the top as a response to locally rising sea level. Many such barrier islands are now highly built up: for example, places such as Atlantic City and Ocean City are built almost literally on the beach. This restricts the natural inland migration.

Cliff erosion is often linked to beach erosion, as removal of protective beaches exacerbates erosion of the cliffs. Cliff erosion is a serious problem along parts of the developed southern California coast, where cliff-top apartment buildings have had to be demolished. Here, cliffs have failed because of tectonic activity coupled with groundwater seepage and wave undercutting of the cliffs. In Britain there have been many instances of buildings collapsing as a result of cliff failure. A recent example was Holbeck Hall hotel in Scarborough on the north-east coast of England. In most such cases, naturally high rates of erosion on failure-prone cliffs have been exacerbated by building which has altered the cliff hydrology.

What causes coastal erosion? Erosion is produced by the interaction of natural and

Table VI.1   Rapid rates of coastal retreat at sites in Britain

Area	Cliff geology	Average retreat rate (metres per 100 years)
North Yorkshire	Glacial drift	28
Holderness	Glacial drift	120
Norfolk (Cromer–Mundesley)	Glacial drift	96
Essex (The Naze)	Glacial drift, London Clay and crag	11–88
Kent, Folkestone	Gault Clay	28
East Sussex, Seaford Head	Chalk	126
East Sussex, Beachy Head	Chalk	106
East Sussex, Cliff End	Sandstone (Hastings Beds)	108
Dorset, Ballard Down	Chalk	23
Dorset, Kimmeridge Bay	Kimmeridge Clay	39

Source: After Goudie (1990, 1995).

Plate VI.2 The jetty at West Bay, Dorset, southern England, has plainly modified the drift of sediment along the coastline. In the foreground sediment has accumulated, but in the background the beach is starved of sediment and erosion is occurring, necessitating coastal protection engineering schemes. (A. S. Goudie)

human factors both acting to increase wave energy and/or reduce sediment availability. The rate of erosion depends upon the interplay between the erosive action of the waves and other agents of erosion, and the erodibility of the rocks and sediments being affected. Natural increases in the tendency of the coastal environment to erode are caused by storms, **El Niño events** and longer-term increases in sea level. All these factors increase wave energy at the coast. Locally, human impacts may be increasing the erodibility of coasts by the following means:

- reducing the availability of sediment for protection, and accelerating erosion by altering the wave energy field and sediment stores with groynes, breakwaters and cliff protection schemes;
- removing vegetation which stabilizes

coastal wetlands, thus making the sediment more erodible;
- reducing the sediment supply by offshore and onshore mining, and by trapping sediment behind dams on rivers that enter the ocean;
- replacing the coastal plain over which barrier islands can migrate with built-up areas, which restrict sediment movements;
- reducing the stability of coastal cliffs through building and altering groundwater levels.

The fact that sediment moves between different parts of the coast means that attempts to reduce erosion in one area can have the opposite effect on areas downdrift. In New Jersey, USA, for example, terminal groynes at Sandy Hook at the southern end of Long Beach Island have encouraged

Plate VI.3   A sea wall and cliff stabilization measures at Weymouth, Dorset, southern England. Such engineering solutions are expensive and are not always successful. (A. S. Goudie)

accelerated erosion downdrift. 'Beach nourishment', that is, 'feeding' the beach by bringing in sediment, has been utilized to overcome such problems, with some success.

It is clear, then, that managing coastal erosion can be a very difficult task. To be successful, it requires understanding both of what factors are causing erosion in a particular area and of how remedial techniques will themselves affect the situation. For example, a highly developed barrier island, where future sea-level rise induced by global warming threatens to exacerbate erosion on a naturally subsiding coast, will require a very different management strategy from a small beach where erosion can be related to a specific episode of offshore sand mining. Clearly, the threat of a future acceleration in sea-level rise because of global warming (see section 2 above) is making coastal erosion an increasingly serious problem. In many places managed retreat, where coastal erosion is allowed to occur relatively naturally, and settlements moved inland, is perhaps the only feasible long-term solution.

FURTHER READING

Nordstrom, K. F., 1994, Developed coasts. In R. W. G. Carter and C. D. Woodroffe (eds), *Coastal Evolution*, 477–509. Cambridge: Cambridge University Press.
A wide-ranging review of the problems faced by coasts with large concentrations of people, in an edited collection of advanced papers.

Bird, E. C. F., 1985, *Coastline Changes: A Global Review*. Chichester: Wiley.
A survey of erosion and accretion on coastlines in various countries.

Erosion at Victoria Beach, Nigeria

Around the port of Lagos is a 200 km long stretch of barrier island coast characterized by a sandy barrier, backed by a mangrove-fringed lagoon. It appears to have grown seawards over the Holocene. Now, however, the coast is eroding at sometimes spectacular rates (Ibe, 1988). Wave energy is high in this environment: the coast is pounded by waves coming all the way across the Atlantic, and there is a general trend from west to east in movement of material along the shore.

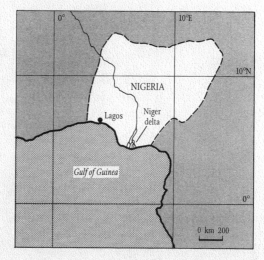

The port and former capital city of Lagos has a population of over 6 million. Much of its economic prosperity is based on the extraction of oil from the Niger Delta. Lagos is situated where there is a break in the coastal barrier, and expanded rapidly in the nineteenth and early twentieth centuries on land reclaimed from mangrove swamps behind the barrier. As the port developed, improvements were made to the harbour, starting with dredging in 1907. Major harbour works began in 1908. These involved the construction of two break-waters and a training wall or jetty to provide a safe, deep-water entry for large ships. These breakwaters interrupted the west-to-east longshore drift. The consequence has been a long-term erosion problem on Victoria Beach (on the west side of the harbour), and accumulation of sand on Lighthouse Beach to the east. Victoria Beach has eroded by up to 69 metres per year since then (by 2 km in all), and an estimated 2.5 sq km of beach has been lost (figure VI.4).

Victoria Beach is an important recreational area for Lagos. Also, its erosion was starting to threaten housing built on low-lying reclaimed land behind the beach, which protects the coast under natural conditions. Beach nourishment by bringing in sediment has been used to try to solve the problem, starting in 1976. Although it was successful in the short term, dramatic erosion occurred in 1980, necessitating further emergency nourishment using 2.1 million cu metres of sand between 1980 and 1981.

The erosion problems of Victoria Beach are particularly hard to solve because the Lagos port must be kept open. The sand accumulating on Lighthouse Beach is also proving to be a problem, as eventually it will extend past the western breakwater and be washed around into the harbour. Long-term, integrated management of the entire coast here is necessary. This may involve pumping sand around from west to east (mimicking the natural longshore drift) and preventing further development on vulnerable low-lying land.

Erosion of Victoria Beach must be set in the context of more general erosional trends along the Nigerian coast. Altogether, Nigeria has some 800 km of coastline and there is much evidence of widespread erosion within the past few decades.

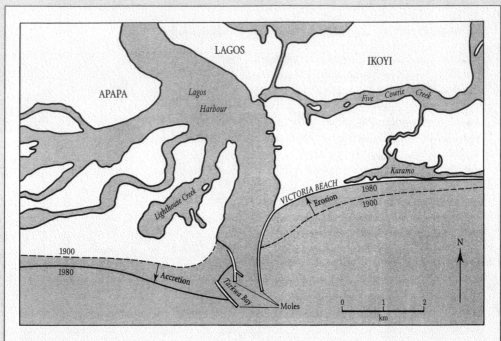

Figure VI.4 Since breakwaters were built, erosion and accretion have resulted along the beaches around Lagos harbour, Nigeria
Source: After Usoro (1985).

Along the Niger Delta coast, which is situated to the east of the Lagos area, erosion – coupled with environmental problems related to the oil extraction industry – is a serious problem and several schemes (usually involving beach nourishment, like that at Victoria Beach) have been implemented with limited success. Table VI.2 below shows some recent typical erosion rates along barrier beaches of the Niger Delta coast for comparison with those at Victoria Beach.

Table VI.2 Erosion on the Niger Delta coast	
Location	*Erosion rate (metres per year)*
Ogboiodo/Escravos (western part of the Niger Delta coast)	18–24
Forcados (western part of the Niger Delta coast)	20–22
Brass (central part of the Niger Delta coast)	16–19
Kulama (central part of the Niger Delta coast)	15–20
Bonny (eastern part of the Niger Delta coast)	20–24
Imo/Opobo (eastern part of the Niger Delta coast)	10–14

4   COASTAL FLOODING

Simply put, coastal flooding is a result of substantially increased water levels on the coastal plain above high tide level. Floods occur, therefore, as the sea level rises, or the coast sinks, or where a combination of the two happens. The possibility of global warming raising sea levels worldwide is making coastal flooding an ever more serious issue. Currently, flooding affects low-lying coasts such as the Mississippi, Nile and Ganges deltas, the Thames estuary, Venice, Bangkok and the Netherlands. In many areas expensive flood protection structures and schemes have been implemented, usually after a serious flood. An example is the Thames Barrier, completed in 1982. In Bangladesh, storm surges produced by cyclones in the Bay of Bengal have produced particularly devastating floods, such as that in April 1991 which is estimated to have killed more than 100,000 perople.

The major causes of coastal flooding are: storm surges; El Niño events; hurricanes; tidal waves (tsunamis); and subsidence, through abrupt tectonic movements. The size and severity of flooding are influenced by the tidal regime and the phase of the tide when the flood event strikes. In estuaries, peak river flows can also make matters worse.

Factors which make places more prone to flooding, by lowering the land, include:

- natural compaction of delta sediments, which promotes subsidence;
- oil, gas and groundwater extraction, which promotes subsidence;
- removal of mangrove and marsh vegetation, which reduces coastal protection for backshore areas;
- building on low-lying, subsiding land;
- failure of flood defences such as dikes.

There are several stages in management of the coastal flood hazard. The initial stages include understanding the major causes of flooding in the area; building structures and flood defence schemes; and improving prediction and disaster planning. In Bangladesh, for example, mangrove trees have been planted on a large scale to encourage the stabilization and development of mangrove swamps. These help to provide a 'buffer' and so to prevent flooding inland. Also, the Coastal Embankment Project has been established to build embankments and a series of sluices to protect against flooding. Flood hazard warning improvements and increased provision of emergency shelters on high land are also vital elements in flood management here.

The management of coastal flooding in Britain is in the hands of the Ministry of Agriculture, Fisheries and Food (MAFF). Since 1985, MAFF has also managed coastal protection works, for example to control erosion. The Environment Agency, formerly the National Rivers Authority, also has an important role to play in flood warning and flood defences. The floods on the east coast in 1953 provided a major stimulus to planning and defences in Britain; in East Anglia most of the sea defence structures date from the decade after 1953. A national network of tide gauges and the Storm Tide Warning System (STWS) was also set up about this time.

FURTHER READING

Perry, A. H., 1981, *Environmental Hazards in the British Isles*. London: Allen and Unwin.

Ward, R. C., 1978, *Floods: A Geographical Perspective*. London: Macmillan.

Flooding at Towyn, North Wales, February 1990

Towyn and the surrounding Clwyd coastal lowlands, covering about 20 sq km altogether, support a population of around 14,000 people. On 26 February 1990 the sea wall at Towyn was breached as a result of a storm surge. The floodwaters rose to over 5 metres above normal sea level or Ordnance Datum (OD) in the centre of Towyn. Over 6.4 sq km of land was flooded, including all of Towyn and much of the adjacent settlement of Kinmel Bay (figure VI.5). Many houses, bungalows and caravans were destroyed. Over 750 domestic and commercial properties were ruined in Towyn alone. The floodwaters reached up to 2 km inland, covering much agricultural land.

What caused these floods, and why were the floodwaters so patchily distributed? Storm surges are a major cause of coastal flooding around the British coast. Their low barometric pressure and strong winds act to raise tide levels above those predicted. When a strong depression occurs over the sea, falling barometric pressure acts to 'suck up' the water surface, producing a rise of about 1 cm for every 1 millibar (mb) drop in pressure.

On 22 February 1990 there was a large anticyclone situated over central Europe and a strong depression over south-west Iceland. Between 23 and 25 February this depression deepened and moved towards southern Scandinavia. By 26 February it was just west of Denmark and a second, related depression had developed to the south-east of Iceland. These depressions had low-pressure cores

Plate VI.4   A flooded caravan and trailer park beind the sea wall at Towyn, North Wales, in March 1990. (Richard Smith/Katz)

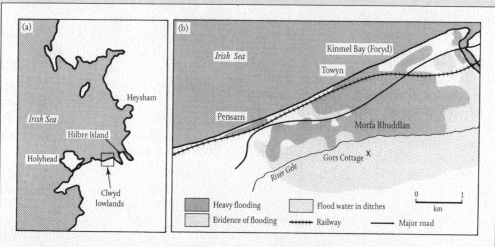

Figure VI.5 (a) Location of Clwyd lowlands; (b) The extent of flooding west of the River Clwyd, near Towyn, North Wales, in February 1990
Source: After Englefield et al. (1990), fig. 3.

of between 950 and 960 mb, and major storms with high winds occurred across the UK between 25 and 26 February. Rain, hail, gale-force winds and low pressure occurred in the Clwyd lowlands, coupled with exceptionally high sea levels (assisted by the storm surge conditions). This combination of circumstances led to the floods.

The Clwyd lowlands were particularly vulnerable to flooding as they are extremely low-lying, situated on reclaimed land ranging from 3.5 metres OD to 7 metres OD, most of it below 5 metres. Along much of the coast there is a natural protective shingle ridge; this reaches a height of up to 7 metres at Kinmel Bay, but dies out at Towyn.

The history of land use and human intervention in the area also had a crucial role to play in the flooding. In 1847 the Chester to Holyhead railway was opened, running along the coast (figure VI.5). This has interfered with coastal sediment movements ever since, resulting in long-term coastal erosion problems. For example, between 1872 and 1899, along one section the coast eroded by 60 metres in front of the railway line. During the nineteenth and early twentieth century sea defences were built to help overcome these problems, including a sea wall and groynes at Towyn. Very little sediment ever accumulated in front of the wall, and on 26 February 1990 it faced the full force of the ocean, breaking at around 11 a.m.

Studies carried out after the flood by Englefield et al. (1990) showed the extent of the damage and explained that the pattern of flooding was controlled by microtopography and the road layout within settlements. Roads and higher areas acted as flood barriers. Interestingly, the old bungalows nearest to the sea at Kinmel Bay escaped the worst flooding, as they were located on the shingle ridge at 6–7 metres OD.

5 COASTAL AND MARINE POLLUTION

A report in 1990 by the Group of Experts on the Scientific Aspects of Marine Pollution concluded that most of the world's coasts are polluted, while many parts of the open ocean are still relatively clean. Coastal pollution is an important environmental issue, affecting human health and the diversity of fisheries and coastal ecosystems. Recent attention has focused on litter and sewage pollution on British beaches; 'red tides' (algal blooms caused by an excess of nutrients; see part IV, section 7) in the Mediterranean and elsewhere; and oil spills such as that resulting from the wreck of the *Braer* tanker off the Shetland Islands in January 1993 and the *Sea Empress* off West Wales in February 1996.

Most marine and coastal pollution (over 75 per cent) comes initially from land-based sources. It is brought down to the sea by rivers, dumped in sewage outfalls, or arrives via the atmosphere. The rest comes from dumping by ships and from offshore mining and oil production. To take oil pollution as an example: a surprising 34.4 per cent of a total of 3.2 million tonnes per year which reaches the sea comes from land, via urban runoff etc.; another 34.3 per cent comes from marine transport (oil shipment); the rest comes from atmospheric fall-out, offshore oil production (only 1.6 per cent) and natural sources.

The major types and sources of coastal pollution are:

- nutrients from sewage, agricultural runoff, aquaculture;
- pathogenic organisms from sewage;
- litter, especially plastics, from land and ships;
- metals, e.g. cadmium and lead, from mining and industry;
- sediments, from deforestation, soil erosion, mining and dredging, which

may be contaminated with synthetic organic compounds;
- **organochloride** pesticides, from agricultural and industrial runoff;
- PCBs (polychlorinated biphenyls) from industry;
- oil, from land and oil tanker discharges;
- radionuclides, from discharges from nuclear reactors and reprocessing plants and natural sources.

The amounts involved can be horrifying. In 1985 at least 450,000 plastic containers were dumped by the world's shipping fleet. The impacts on humans, coastal ecosystems and coastal stability can be devastating. The effects can also be long-lasting. Oil spilt from the Isla Payardi oil refinery in Panama in 1986, for example, came ashore on to a mangrove coast and killed many shellfish, as well as becoming absorbed in mangrove muds. Five years later the oil was recycled as these sediments eroded, and started to threaten nearby coral reefs.

Other pollutants have more immediate and short-lived effects. For example, nutrients which trigger algal blooms cause deoxygenation of the water, killing other species. These algal blooms may also be toxic, poisoning shellfish and affecting human health. Many synthetic organic compounds, such as PCBs, have a sinister long-term effect. They tend to accumulate in living organisms, gradually getting more concentrated as they are passed up the food chain and seriously affecting marine mammals and sea birds.

Some areas of the coastal and marine environment are particularly prone to such pollution. Especially vulnerable are areas where tidal and wave action encourage the concentration of pollutants, and where sediments can act as a sink. Thus sheltered bays, estuaries and coastal lagoons are key areas affected by pollution. Enclosed seas such as the Mediterranean and Baltic are now also seriously affected by pollution over vast areas.

Plate VI.5   Beach pollution at Bahrain, Arabian Gulf. (A. S. Goudie)

What can we do to reduce coastal pollution? Great strides have already been made in limiting the influx of pollutants through a number of international agreements. In 1987, for example, eight countries bordering the North Sea agreed to phase out the incineration of chemical wastes at sea by 1994. Dumping of radioactive waste at sea was stopped worldwide in 1982. The United Nations Environment Programme (UNEP) has coordinated many attempts to tackle pollution in particular areas, such as the south-east Pacific and the Black Sea (which has already suffered serious ecological damage from toxic chemicals, pathogens and eutrophication – see part IV, section 7). Agriculture, industry, urbanization, **mariculture**, marine transport, dumping, oil extraction, mining and war are all important polluters. All must be addressed if the problem is going to be tackled successfully. In many areas, it is still difficult to get accurate information on coastal pollution and its effects. During and after the Gulf War, for example, disagreements raged about how much the war had increased coastal pollution (through deliberate sabotage of oil fields) or decreased it (through preventing oil shipments, and their associated pollution).

FURTHER READING

GESAMP, 1990, *The State of the Marine Environment*. Oxford: Blackwell Scientific. A general report by an authoritative international group on pollution of the oceans and the coastal seas.

Clark, R. B., 1989, *Marine Pollution*, 2nd edn. Oxford: Clarendon Press. A very good overview.

Pollution in the Mediterranean Sea

It is estimated that the population of countries around the Mediterranean will rise to 430 million by 2000 CE. These countries, and especially their coastal zones, also attract large numbers of tourists: 100 million visited the area in 1984. There are huge disparities between the economies of countries on the northern and southern sides of the Mediterranean, but pollution is getting worse everywhere.

The major types of pollution are:

- oil;
- domestic waste;
- industrial and urban wastewater;
- organochlorine pesticides;
- heavy metals;
- PCBs.

All these harm wildlife, affect human health and may lead to long-term damage to the entire Mediterranean ecosystem. Oil pollution is now a chronic problem over most of the Mediterranean as a result of tankers discharging ballast and bilge waters in the network of shipping lanes which criss-cross the sea, carrying some 250 million tonnes of oil per year. Sewage is a severe problem, especially around the Italian, Spanish and French coasts (see figure VI.6). The costs of reducing such sewage pollution may be very high. In 1990, GESAMP suggested it would cost US$150 per person to construct sewage treatment and disposal facilities for all the 132 million inhabitants of the coastal settlements around the Mediterranean. That would amount to US$18 billion overall at 1990 prices. Sewage pollution can make swimmers ill, and can also contaminate seafood. In 1973 a cholera epidemic broke out in Naples, Italy, because of contaminated molluscs, and hepatitis can also be transmitted by seafood. Organochlorine pesticides, PCBs and heavy metals are a problem in particular areas such as the Venice lagoon, where lack of 'flushing' allows them to accumulate in bottom sediments. Sewage leads to algal blooms and red tides under extreme circumstances, which first posed a problem in the Gulf of Venice in 1972. Eutrophication is a serious problem in the western Adriatic Sea, where rivers coming from Italy bring around 29,000 cu metres of phosphates and over 120,000 cu metres of nitrates every year. Like acid rain, this results in a transnational problem: beaches along the coast of Croatia are affected as seriously as Italian beaches.

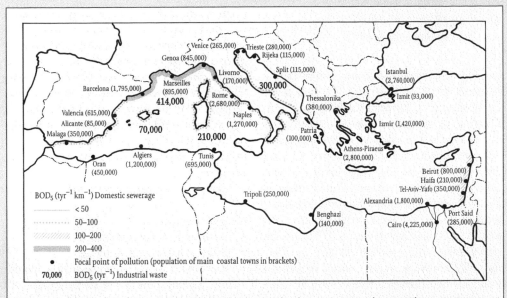

Figure VI.6 Sewage and industrial waste discharges into the Mediterranean Sea (BOD = biological oxygen demand)
Source: After Clark (1989), fig. 9.3.

Some pollution problems affect only small areas and are easily solved. An example is the discharge of tannery wastes contaminated with chromium into the Gulf of Geras on the Isle of Lesbos. The impact of this was lessened in 1983 when an effluent treatment plant was installed (Papathanassiou and Zenelos, 1993). Some pollution problems, however, are less easily solved; and some areas, such as the Venice lagoon, appear to be polluted beyond acceptable limits.

In 1979, as a response to concerns about many of these issues, the 'Blue Plan', set up with the help of UNEP, was adopted by the Mediterranean countries. This plan aimed to help both economic development and environmental protection, and to limit pollution from land-based sources. As always, however, dealing with international environmental problems is a difficult task, and implementing sustainable development and tourism has, so far, proved very hard. As tourism is one of the major industries of the Mediterranean, and is affected by pollution as well as contributing to it, perhaps the initiation of ecotourism would make a start towards solving the pollution problems here.

6   COASTAL DUNE MANAGEMENT

Coastal dunes provide an important buffer between land and sea, and act as a store for sediment. They are closely linked with beaches, as there is a regular interchange of sediment, nutrients and organisms between beach and dune systems. Coastal dunes are a common component of most coastlines, and are often of very impressive height and extent. Notable examples are found on the west coast of America, where the Coos Bay dunes are 72 km long and reach heights of 50 metres. In Europe, high dunes occur along the Coto Doñana in southern Spain, where they reach 90 metres. Coastal dunes, unlike many desert dunes, tend to be vegetated. Hardy salt-tolerant plants grow on them nearest the sea; as dune environments get more sheltered and better soils occur further inland and over time, other plants follow.

Coastal dunes provide many attractions for human society. Because of this, as well as their natural dynamism and role as agents of coastal protection, their successful management has become an important issue, especially as most sandy coastlines are undergoing erosion. As cliffs are prevented from eroding, so the supply of eroded material going to beaches and dunes is reduced. In Britain, for example, dunes on the East Anglian coast probably now have a diminished supply of sand because of coastal protection works covering about 60 per cent of the coast here. Dunes themselves are eroded by both wind and waves.

Sand dunes provide a harsh environment, colonized at first by hardy plants that can tolerate salt and sand, such as sea rocket (*Cakile maritima*) and salt wort (*Salsola kali*), whose seeds can tolerate long periods soaked in seawater. As these plants grow, they trap sand and help the dunes to grow. Grasses, such as *Ammophilia arenaria* (marram grass) and sand couch-grass (*Agropyron junceiforme*), are the major sand-accumulating species. Gradually, plant succession creates a diverse ecosystem, which is attractive to birds, insects, reptiles and small mammals. For example, half the flowering plants of Britain can be found in coastal dune areas around the country.

Important human uses of dunes include:

• golf courses;
• sand and water extraction;
• afforestation and grazing;
• recreation, such as horse-riding, walking, biking and off-road vehicles;
• military training and exercises;
• housing, camping and caravan parks;
• transport, such as roads and airfields;
• pipelines.

Most of these uses, however, involve disturbing the natural ecosystem. Such disturbance often encourages dune mobilization and destabilization and the development of **blowouts**. This can lead to sand migrating inland over valuable agricultural land or housing; it also removes the coastal protection afforded by the dunes. Other impacts affect the groundwater level of dunes, which in turn affects the ecology. In the Netherlands, for example, coastal dunes provide an important source of drinking water. Other human uses of dunes can 'fossilize' them, removing any chance of natural dynamism through such things as planting grass and trees for golf courses. Finally, some human impacts affect dunes indirectly: removing sand from beaches, damming rivers, offshore sand mining and pollution can all tip the balance between sedimentation and erosion.

Because of the many and various uses and abuses of coastal dunes, considerable money and time have been invested in trying to conserve and protect dunes in order to save them and the wildlife they support. Dune management schemes usually involve all or some of the following:

- aiding deposition of sand on beaches, through groynes, sea walls and beach nourishment;
- shaping dunes, using bulldozers to move sand;
- planting and watering dunes;
- using **biofabrics**, mulches, etc. to help stabilize fragile dune surfaces;
- fencing to restrict access;
- providing walkways to channel people away from sensitive areas and prevent damage to the underlying dune;

- providing signs, information displays and education to involve the public in dune conservation.

However, overmanagement can also be a serious problem. Most coastal dunes under natural circumstances are not fixed, and movement of dunes and blowouts are perfectly natural occurrences. Some element of disturbance needs to be included in successful dune management schemes. Figure VI.7 shows how dune management is also affected by natural coastal erosion: on the Baltic coast of Poland a catastrophic storm in January 1983 led to severe erosion of beach and foredunes, which then threatened the stabilized dunes behind (Piotrowska, 1989).

Plate VI.6   Footpaths causing erosion patterns across coastal sand dunes at Winterton, Norfolk, eastern England. This is a clear example of the effects of recreational pressures on the landscape. (University of Cambridge Air Photograph collection)

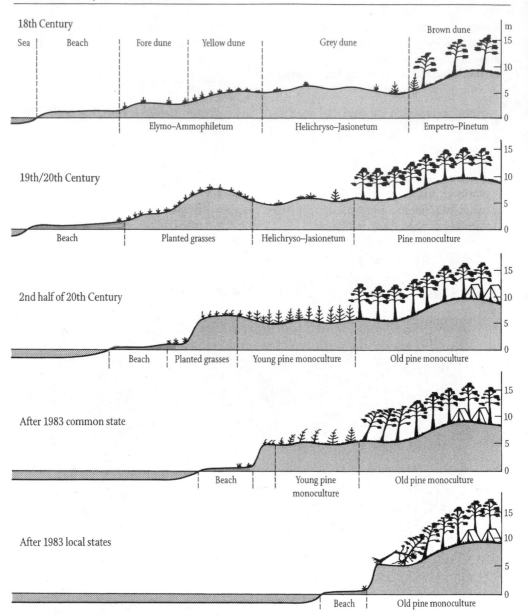

Figure VI.7   The history of dune management and coastal erosion on the Baltic coast of Poland
Source: After Piotrowska (1989), fig. 5.

FURTHER READING

Ranwell, D. S. and Boar, R., 1986, *Coast Dune Management Guide*. Huntingdon: Institute of Terrestrial Ecology.
A useful, practical guide to management techniques, with plenty of case studies.

Managing dunes on the Lancashire coast, England

The Ainsdale–Formby dunes cover 800 hectares, of which 490 hectares is a National Nature Reserve. The coast here faces north-west and the dunes form a suite of more or less parallel ridges with low-lying areas called dune slacks in between. Behind these are more irregular dunes. Erosion has dominated at the south end of the area since the beginning of the twentieth century. The coast is protected in the north towards Stockport by wide sand flats. The sand here is rich in calcium carbonate, so dune soils are not very acidified. The area is rich in plant species, with marram, dune scrub and woodland including *Alnus* and *Betula*. *Hippophae rhamnoides* (sea buckthorn) was introduced here and has spread considerably over the dunes (Boorman, 1993). In 1959 myxamatosis arrived, decimating the rabbit population and aiding the spread of scrub (by preventing rabbit grazing which maintains grass). Now, the Formby dunes to the south of the nature reserve are threatened by both public pressure and coastal erosion.

Detailed studies by the geomorphologist Ken Pye have revealed the long-term importance of human activities to the dunes here (Pye, 1990). Marram grass was introduced into the area at the start of the eighteenth century, when strict laws were introduced to encourage planting; indeed, planting was obligatory until 1866. Marram favours the development of hummocky sand hills, as found towards the back of today's dune system. In the late nineteenth and early twentieth century, brushwood fencing and backshore planting were used to encourage dune development, resulting in the parallel dune ridges found over most of the dune system.

By the late 1920s recreational pressure was causing severe erosion and producing much blown sand. Other activities which have affected these dunes include:

- excavation of flat-floored depressions for asparagus cultivation;
- sand mining;
- waste dumping;
- development of caravan and car parks;
- road-building;
- development of golf courses.

These pressures have led to dune management and restoration schemes, such as the Sefton Coast Management Scheme, established in 1978. This scheme began to restore the dunes using brushwood fencing, marram planting and wooden fencing, and restricting access by vehicles and pedestrians. Covering some 17 km of coast, the Sefton Coast Management Scheme provides a framework for nature conservation projects within an area which includes several different landowners. A Coast Management Officer has been appointed who promotes co-operation between the different landowners and ensures integrated management of this sensitive coastal environment.

7   CORAL REEF DEGRADATION

Coral reefs are some of the world's most diverse ecosystems, containing a bewildering array of corals, fish and other organisms. Although they cover only 0.17 per cent of the ocean floor (an area roughly the size of Texas), they are home to perhaps 25 per cent of all marine species. One hundred and nine countries have between them over 100,000 km of reefs, and many of these are threatened by a series of natural and human-induced stresses (figure VI.8).

Coral reefs require very specific environmental conditions. Reef-forming corals only grow in waters with temperatures of 25–29°C; where there is a suitable, relatively shallow platform less than 100 metres below sea level to grow on; and where sediment and pollution do not kill them off. Thus, their growth is restricted to suitable tropical and subtropical shores. One such is the north-eastern coast of Australia, where the Great Barrier Reef forms the largest agglomeration of reefs in the world, stretching for over 2,000 km and comprising over 2,500 individual reefs. Other major reefs are found along the Gulf coast of Belize and around many South Pacific islands.

As Charles Darwin explained in the nineteenth century, there are three main types of reefs, related in a genetic sequence. First, there are fringing reefs, which connect directly with the shore. Then there are barrier reefs, which are separated from the shore by a lagoon. Finally, when such reefs are growing around a gently subsiding oceanic island, **atolls** are produced. An atoll is a ring of coral reefs around a lagoon, in the centre of which was once the island. Sand and gravel islands accumulating on the margins of such atolls provide a precarious home for flora, fauna and humans, as in the Maldives.

Reefs are remarkable in that their entire geological structure is formed from biological growths, now dead, covered by a thin veneer of living corals. Despite the name 'coral reefs', most reefs are in fact composed of a number of important reef-building species, including coralline algae as well as a range of corals.

Stresses affecting coral reefs in today's world include:

- storms and hurricanes;
- El Niño events;
- sea-level rise and other effects of global warming;
- outbreaks of disease and predators (such as the Crown of Thorns starfish);
- increased sedimentation produced by deforestation on land;
- eutrophication produced by sewage and other pollutants;

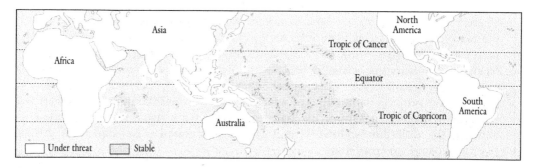

Figure VI.8 A generalized map of threatened coral reefs around the world
Source: After C. R. Wilkinson, personal communication/University of Guam.

- onshore and offshore mining, producing sediment enriched with heavy metals;
- trampling and physical damage from boats and divers;
- overfishing and the use of damaging fishing techniques such as dynamiting;
- direct quarrying and removal of corals for building or curios;
- oil pollution from land and shipping;
- nuclear weapons testing and other military activity;
- pollution and damage from landfill (used, for example, to create new land for airports, and sometimes constructed with toxic waste).

Natural disturbances such as hurricanes can damage fragile corals and fling them up on to the reef flat. However, the impact of such events is probably short-lived and may, in fact, be good for the overall health of reefs, providing a disturbance which may increase species diversity and growth in the long term. El Niño events, which occur every two to ten years and involve widespread changes in ocean currents and temperature, have a potentially more serious effect. They temporarily warm the water around reefs: this can cause coral bleaching, when the corals expel the zooxanthellae, the tiny algae that live **symbiotically** with them. In severe cases, bleaching can cause mass death of corals. Global warming may make such bleaching episodes more frequent and more serious, as it will heat the oceans, and may provide further stresses by accelerating the rate of sea-level rise, forcing corals to grow faster in order to keep up with sea level. Locally some corals have been badly affected by outbreaks of pests and diseases. Crown of Thorns starfish, for example, eat corals; these predators spread rapidly across many South Pacific reefs in the 1960s. In 1993 South Pacific reefs were first observed to be suffering from another biological problem: CLOD (coralline lethal orange disease), which affects coralline algae, another important part of reef frameworks (Littler and Littler, 1995). The causes of such biological disturbances are unknown and much-debated, but they may be at least partly due to environmental pollution.

Other stresses on coral reefs can clearly be blamed on human impacts, both directly on the reefs themselves (from diving and fishing, for example) or indirectly from activities on land or offshore. Increased sediment load, pollution from sewage, agriculture and industry, and destructive fishing techniques all damage the reef ecosystem by upsetting the balance of species. At the Green Island resort on the Great Barrier Reef, sewage has led to an increase in the area of seagrasses, largely at the expense of corals. These seagrasses trap sediments which usually circulate freely around the beaches of the island. Thus pollution here is damaging both reef and beach environments. Deforestation in Thailand and conversion of forest to rubber and cocoa plantations has had severe impacts on the reefs on southern Phuket Island, producing excessive sedimentation and killing corals.

Many of these stresses on reefs are now acting together, and many reefs are going into the twenty-first century in an increasingly unhealthy state (table VI.3). If global warming continues, some reefs may be unable to cope. How serious is the problem and what can we do about it?

Reefs have many uses and roles for society:

- they are agents of coastal protection, providing a natural 'sponge' absorbing wave energy;
- they are major tourist attractions;
- they are an important focus for biodiversity and conservation of marine species;
- they contain living and non-living resources of great economic value, such as fish, crustaceans, coral rock and sediment.

Plate VI.7   The destruction of a coral reef by draglines used to build a new port on the island of Taketoni off Okinawa, Japan. (Panos Pictures/Jim Holmes)

Most countries cannot afford to lose their reefs. In terms of fishing alone, Pacific islanders get up to 90 per cent of their protein requirements from reef fish, and worldwide reefs are home to a total fish catch of 4–8 million tonnes per year (Weber, 1993).

Future sea-level rise will affect reefs as reef-building corals and algae only grow within relatively shallow water. Three major reef strategies have been identified (figure VI.9): 'Keep-up', 'Catch-up' or 'Give-up', depending on the balance between the relative rate of sea-level rise and the growth rate of the corals involved. If sea levels rise very fast, most reefs will be unable to keep up. Given the recent predictions of 4–5 mm per year mean sea-level rise over the next 50–100 years (see section 2 above), most reefs will keep up or catch up. Unhealthy reefs, however, are less able to keep their growth rates up and are more likely to give up.

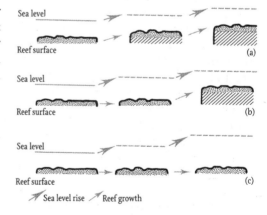

Figure VI.9   Coral reef growth scenarios: (a) 'keep-up': reef growth proceeds at roughly the same rate as sea-level rise; (b) 'catch-up': sea level initially rises faster than reef growth; then reef growth catches up; (c) 'give-up': sea level rises too fast for slow-growing or unhealthy coral reefs

Table VI.3 Summary of the health of coral reefs in various parts of the world

Area	% world's reefs found in the area	Reef health
South-east Asia	30	60–70% reefs sick. Deforestation, mining and fishing problems
Pacific Ocean	25	Mainly good condition, because of low populations
Indian Ocean	24	20% reefs lost. Mining, fishing and coastal pollution problems
Caribbean Sea	8	Deforestation and tourism problems
Atlantic Ocean	6	Coastal development and tourism problems. Bermuda has good reef reserves
Middle East	6	Low runoff, low population and little tourism aid reef health; oil spills a problem

Source: Adapted from Weber (1993), table 3–2.

Currently, many reefs are protected to varying degrees in an attempt to reduce the stresses on them. The Great Barrier Reef Marine Park in Australia was created in the 1970s. It contains five sections, with different reef uses allowed in each. Oil drilling and mining are prohibited throughout the park, and in some sections only scientific research and traditional fishing are permitted. Where relatively poor countries have vulnerable reefs which are also major tourist attractions, there can be many conflicts involved in successful reef management, and marine parks can be hard to monitor and control. If reef management and protection is to be successful, it is necessary to understand how reefs work, manage the various human uses of them, and plan onshore land use to reduce damage from external sources. The problem has many dimensions, as Weber (1993, p. 53) explains: 'Ultimately, the forces behind reef decline are hard to untangle. Overexploitation and coastal pollution stem from business interests, wealthy consumers, the growing numbers of coastal poor and governments trying to balance conflicting development goals. No single group is the cause of reefs' precipitous decline, yet all contribute to the tragedy.'

FURTHER READING

Guilcher, A., 1988, *Coral Reef Geomorphology*. Chichester: Wiley.
A general, clearly written study of coral reefs with a useful section on human pressures on reefs.

Threatened reefs of the Red Sea

The Red Sea, which extends from 13° N to 30° N, has fringing reefs along almost all of its coastline. Reefs are especially well developed along the north and central coasts. Conditions are particularly suitable for reef growth here. There are no permanent rivers flowing into the Red Sea from its arid hinterland and phytoplankton productivity is low, both of which encourage clear water. There are few storms and no tropical cyclones. Reefs along the northernmost part of the Red Sea, however, are affected by occasional extremely low tides, and sea temperatures here are near the minimum level acceptable for reef-building corals. Most countries bordering on the Red Sea are arid and sparsely populated, and there have therefore been few onshore threats to the reefs. Pollution from the busy Red Sea shipping lanes is a problem, and oil pollution from oil exploration in the Gulf of Suez is especially serious.

Tourism is a growth industry here, most concentrated in the northern countries of Israel, Egypt and Jordan. Studies estimate that 19 per cent of Egypt's reefs are now affected by tourism, and this figure is expected to rise to 73 per cent by the year 2000. The Egyptian resort of Hurghada provides a good example of the actual and potential impacts of tourism on Red Sea reefs. The town of Hurghada was founded in 1909 to supply the oil industry. It did not start to attract many tourists until the late 1970s. Now it has huge tourist complexes stretching some 20 km along the coast and many more are planned (figure VI.10). Diving is a major attraction for tourists here. What damage does tourism cause to these reefs?

First, construction creates dust which, in the dry Red Sea climate, gets blown on to reefs, creating a sediment nuisance. Secondly, construction often involves creation of new coastal land from landfill. This can cause major damage to reefs. Also, enhancement of tourist beaches through beach nourishment, etc., can upset regional sediment dynamics. Thirdly, sewage disposal, desalination, irrigation and rubbish disposal all pose problems. At Hurghada, sewage is treated before it enters the sea, but observations of high algal growth on reefs nearest the shore suggest that high nutrient inputs may still be a problem (Hawkins and Roberts, 1994). Fourthly, tourism may encourage overfishing and the collection of corals and shells for sale. Fifthly, diving and boat anchoring have been shown to damage reefs over small areas. Finally, it should be noted that tourism has positive benefits for neighbouring reefs, as it reduces the industrial development in the area and brings an added awareness of the value of natural reef habitats.

A Marine Station was established at Hurghada in 1931, which has provided invaluable data on marine biology. A national park has been proposed to help protect the reefs. Oil pollution remains a serious problem. For example, in 1982 fresh oil was found over a wide area, affecting turtles, white shark, spoonbill and osprey (Wells, 1988).

Further north, around the tourist resort of Sharm-el-Sheikh, the Ras Moham-med Marine Park was set up in 1983 to aid reef conservation. Here, there is a

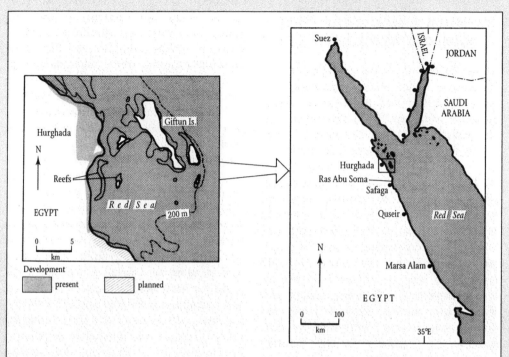

Figure VI.10   Present and planned coastal tourist development around Hurghada, Egypt
Source: After Hawkins et al. (1993).

high density and diversity of corals, as well as sharks, giant clams, green turtles and many interesting bird species.

According to a recent study, tourism is causing 'worrying rather than alarming' damage to Red Sea reefs. However, the situation could easily worsen as tourist numbers grow and global warming and natural stresses compound the problems. Natural stresses include outbreaks of sea urchins and other grazing organisms. Sea urchins can reach high population densities on the coral reefs here. They graze on coral and can inflict damage on the reefs. Several areas of reefs along the Red Sea coast have shown signs of urchin damage in the past, and similar problems may recur in the future.

Further reading

Hawkins, J. P. and Roberts, C. M., 1994, The growth of coastal tourism in the Red Sea: present and future effects on coral reefs. *Ambio* 23, 503–8.

8 AQUACULTURE AND COASTAL WETLANDS

Aquaculture is the water-based version of agriculture, where plants and animals are grown and harvested for food and other products. Since the 1970s aquaculture has developed enormously, and now accounts for about 15 million tonnes or 17 per cent of world fisheries production (figure VI.11). Aquaculture can take place inland, on freshwater lakes and ponds, but a large proportion of aquaculture takes place in brackish water or seawater ponds in coastal wetlands. Along tropical coasts, for example, it is estimated that about 765,000 hectares of land are currently in use for shrimp production. Shrimps, oysters, catfish, tilapia, salmon, rainbow trout and tiger prawns, among a wide range of other species, are regularly farmed through aquacultural techniques.

Why are coastal wetlands commonly converted to aquacultural use? And why does it matter? Coastal wetlands, which include salt marshes, mangrove swamps and mud flats, are found along low-lying, sheltered coastlines with a large sediment supply. In general, they form in environ-

ments where wave energy is low, and where tidal processes dominate, such as estuaries, deltas and bays. In the upper intertidal zone and above, salt-tolerant vegetation may grow. In the temperate zone, salt marsh communities such as *Spartina* grasses dominate, grading into mangrove trees (e.g. species of *Rhizophora* and *Avicennia*) in the tropics. At lower tidal levels there are mudflat surfaces which look bare, but actually support large numbers of algae and mud-dwelling animals.

Coastal wetlands have often been seen as 'wastelands' but, like other wetlands (see part II, section 9), they play some very useful roles. These include acting as a natural agent of coastal protection, buffering the land behind them from the sea, and acting as a purifying agent, by removing toxic wastes from the water entering them. They are also invaluable in preserving biodiversity: for example, they provide important stopping-off points for numerous migrating birds. In mangrove swamps, the mangrove trees themselves are a useful source of timber and firewood for many local communities.

There are many large areas of coastal wetlands, such as nearly 600,000 hectares of salt marsh on the Atlantic coast of the USA, and an estimated 22 million hectares of mangrove swamps worldwide. Many coastal wetlands are threatened by development. Agriculture, industry and urban expansions can all lead to land reclamation and the removal of natural mangrove ecosystems. Aquaculture also leads to disruption of the natural coastal wetland as trees and other natural vegetation are cleared, ponds dug and filled with water, and nutrients and waste products discharged into the water. Eutrophication can become a problem as a result of the influx of nutrients. The species mix may be affected and total biodiversity reduced.

In Indonesia, for example, brackish water fishponds (locally known as *tambak*) now occupy over 269,000 hectares or 6.5 per

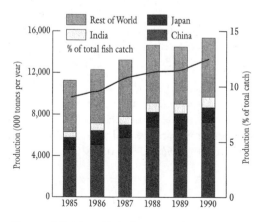

Figure VI.11   Global aquaculture production, 1985–1990
Source: After UNEP (1993), fig. 3.10.

Plate VI.8   Aquaculture is expanding rapidly in South-East Asia. These fish ponds are located on Java, Indonesia. The creation of fish ponds can destroy important natural coastal vegetation and contribute to coastal pollution. (Panos Pictures/ Jeremy Hartley)

cent of the total former mangrove area. Conversion to *tambak* is often unsuccessful, as erosion and pollution can become serious problems if the sites are not chosen correctly. As with agriculture on land, aquaculture will only succeed in the long term, without causing ecological damage, if there is a good understanding of how the natural environment works and aquacultural techniques are developed that avoid disturbing these environmental systems too much. In the Far East, where aquaculture has been practised for thousands of years, technological improvements and more sensitive management techniques are helping to reduce environmental problems associated with aquaculture. Technological improvements include better disease control and nutrition and genetic enhancement. Technological advances also enable mangroves to be planted on dikes around ponds. The mangroves provide useful fuel-wood and fertilizer (from decaying leaves), and protect the ponds from erosion. More sensitive management techniques involve ensuring that aquaculture ponds and mangrove forests are not seen as mutually exclusive.

FURTHER READING

Beveridge, M. C. M., Ross, L. G. and Kelly, L. A., 1994, Aquaculture and biodiversity. *Ambio* 23, 497–503.
An introductory review in a journal that is full of important case studies on many topics covered in this book.

Pond culture in the Philippines

The Philippines consist of some 7,100 islands in all. Between 1920 and 1990 the area of mangroves around these islands shrank from 450,000 hectares to 132,500 hectares. Over the same period the area covered by ponds increased to 223,000 hectares. Around 50 per cent of mangrove loss in the Philippines can be ascribed to the construction of brackish water ponds. By 1991 27 per cent of the total Philippines' fish production (some 26 million tonnes) came from such aquaculture.

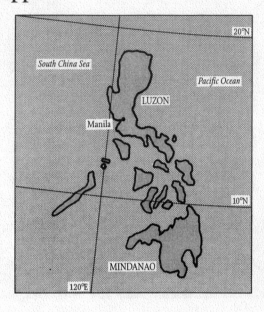

Brackish water pond aquaculture in south-east Asia started in Java, Indonesia in the fifteenth century and spread to the Philippines, where ponds were first constructed on the shores around Manila Bay (Primavera, 1995). There have been several phases of brackish water aquaculture in the Philippines and several effects:

- In the 1950s and 1960s, the government sponsored fishpond development, especially for milkfish production for local consumption;
- The 1970s was declared a conservation decade; and
- The 1980s saw 'shrimp fever', with a boom in production of shrimps, and especially tiger prawns, mainly for export and the urban market.

The notable effects of brackish water pond aquaculture in the Philippines have been mangrove loss; pollution of coastal waters; and decline in production of domestic food crops.

The loss of mangroves affects coastal stability, removes protection against the 20 or so typhoons which affect the Philippines each year, and removes some very versatile plants. There are 26 mangrove tree species found here, many of which have a wide range of traditional uses. The most seriously affected areas are western Visayas and central Luzon.

The ecological damage inflicted by pond aquaculture has prompted the Philippines government and others to take action. Reforestation has been carried out, for example in 1984 when 650 hectares in central Visayas were replanted. As of 1990, 8,705 hectares of mangroves have been successfully planted.

9   CONCLUSION

The world's coastlines and their immediate hinterlands are the focus of a great deal of human activity. They are thus under severe pressure from humankind. By contrast the world's oceans, which are enormous, have so far been much less affected by anthropogenic changes. Their sheer size offers them some protection from the effects of pollution and waste disposal. However, the depletion of world fish stocks is an increasingly serious issue. Halfway between coastlines and the great oceans are the marginal seas – water bodies like the Mediterranean, the Baltic and the North Sea. These do show the clear impacts of a wide range of human activities.

The world's coastlines are experiencing slowly rising sea levels. (There are some exceptions, such as those areas undergoing rapid uplift because of isostatic response or tectonic activity.) If the enhanced greenhouse effect causes global warming to take place, the rate of sea-level rise will increase over the coming decades. Many of the world's coastlines are also being subjected to accelerated rates of erosion or retreat because of a range of human impacts. Some are also being flooded more often, partly because of sea-level rise, but also because of a combination of local human and natural stresses.

Many types of coastal terrain are both dynamic and fragile. Dunes, deltas, beaches, reefs, swamps and marshes come into this category. They all offer many 'ecological services' to humankind. For example, they act as agents of coastal defence or as highly productive ecosystems. Thus we need to treat them with particular care and respect.

Overall, the issues covered in this part of the book illustrate that there are a whole range of immediate environmental problems affecting many parts of the world's coastline, resulting from a combination of human and natural stresses. Future sea-level rises, if they do occur, will be affecting coastlines which are already stressed, and therefore unlikely to be able to respond as they would naturally to such changes. Furthermore, as we have shown, coasts are naturally dynamic over a range of time-scales and any attempts at coastal management must take this into account. We cannot fossilize the coast. Because of the many attractions of coastal environments, a multitude of people are involved in a wide range of activities within the coastal zone. Effective coastal zone management must involve and consider these people. Finally, several of the examples we have used show the many links between coastal environments and those on land and in the oceans. There are also many links between different segments of the coastline, and between the different components of the coastal enviroment (ecology, sediments, water). In order to cope successfully with all these components and interlinkages, coastal zone management schemes must be truly integrated programmes.

KEY TERMS AND CONCEPTS

aquaculture
coral bleaching
coral reefs
El Niño
eustatic change
isostatic change

salt marshes
sea level
sediment circulation cells
surges
tectonics

Points for Review

Why are coastal areas being placed under increasing pressure?

Why might sea levels rise in some areas in coming decades?

Why are so many stretches of the world's coastlines showing signs of erosion?

How would you aim to reduce the impacts of coastal flooding?

What marine environments are especially prone to the effects of pollution?

Which coastal types do you think are especially fragile and dynamic?

Why should we aim to conserve coastal wetlands and coral reefs?

PART VII

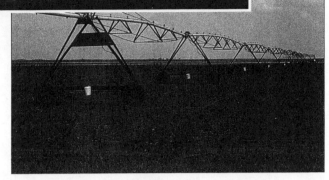

Conclusion

1   INTRODUCTION

The human transformation of nature has been going on for a very long time and has been very pervasive. The Earth's surface still has areas of some size which show little obvious manifestation of the impact of humans (e.g. the deep oceans, parts of the polar regions, some of the tropical rain forests), and we talk of 'wilderness areas' in which very little human activity occurs. However, there is no place on the face of the Earth which is not to some extent affected by the changes in the chemical composition of the atmosphere and associated changes in climate and levels of pollution.

2   THE COMPLEXITY OF THE HUMAN IMPACT

We have demonstrated in this book that different types of human activity cause different types of land transformation. For example, at the one extreme we have discussed some of the changes in the environment that have been caused in and around cities by the process of urbanization. At the other we have demonstrated how even hunters and gatherers living in scattered groups have contributed to such processes as deforestation and desertification. We have selected our case studies to illustrate this theme. We have shown how some changes in the environment are made deliberately by humans but also how many others are accidental by-products of human activity. Often, it takes some time for the environment's responses to such impacts to become apparent. Often, too, their exact causes are hard to identify. In many cases human impacts are increasingly becoming interlinked, and accompanied by natural fluctuations, to produce massive and often unpredictable changes in the environment. We have identified a whole spectrum of different types

of environmental response to stress. These range from short-term fluctuations which can be easily reversed to long-term, potentially irreversible changes which pose intractable problems for environmental management. Our case studies have also illustrated the wide variety of types of attempted solutions to environmental problems. These range from technological 'quick-fix' solutions, such as engineering structures to control coastal erosion, to 'softer' and more 'ecologically friendly' interventions, such as replanting riparian buffer zones to lessen the amount of nitrate pollution that enters rivers from agricultural slopes. Increasingly, any such schemes need to be integrated. That is, environmental problems should not be considered in isolation, but should be viewed as linked parts of the same series of problems. Inevitably, any such schemes will work only if the additional complexities of human society, economy, culture and politics are also taken into account.

3   TOWARDS A SUSTAINABLE FUTURE

It is likely that in coming decades many of the transformations we have described and discussed will become even more important, and the need for effective environmental management even more pressing. Human population levels are increasing, new technologies are emerging, and ever-increasing quantities of energy and resources are being produced and consumed, especially in countries that aspire to the levels of development achieved by some of the world's richest nations. There must be severe doubts as to whether these trends are sustainable. Will the world be transformed by global warming? Will we cut down all our rain forests? Will a large proportion of the world's flora and fauna become extinct? Will many of the world's drylands turn into dust bowls? Will urban

Table VII.1 Some potential adverse impacts of global warming on resources	
Resource	*Possible effects*
Agriculture	Lower crop yields Spread of pests Soil erosion
Forests	Change in rate of growth Change in species composition Shifts in geographical distribution
Conservation areas and nature reserves	Disruption or loss of habitat Invasion of new species
Coastal areas	Inundation of land and accelerated erosion by rising sea levels
Fisheries	Changes in composition of stocks and their location
Water resources	Droughts, floods, changes in amount of supply
Human health	Heat stress Shifts in prevalence of infectious diseases
Energy demand and production	Increases in need for summer air-conditioning

atmospheres continue to become more polluted and more health-threatening? Will our water supplies dwindle in quantity and deteriorate in quality? These are some of the many questions that we can ask about the future. They form the basis of much of the environmental concern that is developing throughout the world.

Are such massive and unwelcome transformations of the face of the Earth inevitable? Is human life sustainable? Can human energies be harnessed over the coming decades to improve rather than to degrade the environment? We are not without hope. We have indicated in many of our case studies that there are ways, means and opportunities to overcome some of the undesirable processes that we have identified. Each and every one of us, in our daily life, has the power to make sure that the generations to come have a sustainable future.

There is now very great interest in how we might adapt to global warming, should it occur. Such adaptations would be necessary if we could not limit emissions of greenhouse gases sufficiently to rule out the possibility of significant warming. They would also be necessary because of the very great range of environments, activities and resources that might be modified as a result of global warming (table VII.1). It is often said that there are two types of adaptation that may be necessary. The first of these is *reactive adaptation*, whereby we respond to climatic change after it occurs. The second is *anticipatory adaptation*, in which we take steps in advance of climatic change to minimize any potentially negative effects or to increase our ability to adapt to changes rapidly and inexpensively.

Reactive adaptation may well be feasible and effective. In many parts of the world

Table VII.2   Examples of 'no-regrets' policies in response to possible global warming

Policy area and measures	Benefits
Coastal zone management	
Wetland preservation and migrations	Maintains healthy wetlands which are more likely to have higher value than artificially created replacements. Maintains existing coastal fisheries that are difficult to relocate
Integrated development of coastal datasets	Integrated data allow formation of comprehensive planning and identification of regions most likely to be affected by physical or social changes. Allows effects of changes to be examined beyond the local or regional scale
Improved development of coastal models	Improved modelling allows more accurate evaluation of how coastal systems respond to climate change and also to other shocks
Land-use planning	Sensible land-use planning, such as the use of land setbacks to control shoreline development, better preserves the landscape and also minimizes the concerns of beach erosion from any cause
Water resources	
Conservation	Reducing demand can increase excess supply, giving more safety margin for future droughts. Using efficient technologies such as drip irrigation reduces demand to some extent. Preserving some flexibility of demand is useful as less valuable uses allow reduced demand during droughts.
Market allocation	Market-based allocation allows water to be diverted to its most efficient uses, in contrast with non-market mechanisms that can result in wasteful uses. Market allocations are able to respond more rapidly to changing supply conditions, and also tend to lower demand, conserving water
Pollution control	Improving water quality by improving the quality of incoming emissions provides greater water quality safety margins during droughts and makes water supply systems less vulnerable to declines in quality because of climate change

Table continues overleaf

Table VII.2 Continued

Policy area and measures	Benefits
River basin planning	Comprehensive planning across a river basin can allow for imposition of cost-effective solutions to water quality and water supply problems. Planning can also help cope with population growth and changes in supply and demand from many causes, including climate change
Drought contingency planning	Plans for short-term measures to adapt to droughts. These measures would help offset droughts of known or greater intensity and duration
Human health	
Weather/health watch warning systems	Warning systems to notify people of heat stress conditions or other dangerous weather situations will allow people to take necessary precautions. This can reduce heat stress and other types of fatalities both now and if heat waves become more severe
Improved public health and pest management procedures	Many diseases which will spread if climate changes are curable or controllable, and efforts in these areas will raise the quality of human life both now and if climate change occurs
Improved surveillance systems	More and better data on the incidence and spread of diseases are necessary to better determine the future paterns of infection and disease spread. This information is helpful under any scenario
Ecosystems	
Protect biodiversity and nature	Biodiversity protection maintains ecological diversity, and richness preserves variety in genotypes for medical and other research. A more diverse gene pool provides more candidates for successful adaption to climate change. One possibility is to preserve endangered species outside of their natural habitat, such as in zoos

Table continues opposite

Table VII.2 Continued

Policy area and measures	Benefits
Protect and enhance migration corridors	Such policies help maintain an ecosystem and animal and tree species diversity. Corridors and buffer zones around current reserve areas that include different altitudes and ecosystems are more likely to withstand climate change by increasing the likelihood of successful animal and tree migration
Watershed protection	Forest cover provides watershed protection, including protection from bank erosion, siltation, and soil losss. All of these functions are extremely valuable whether climate changes occur or not
Agriculture	
Irrigation efficiency	Many improvements are possible and efficient from a cost–benefit standpoint. Improvements allow greater flexibility to future change by reducing water consumption without reducing crop yields
Development of new crop types	Development of more and better heat- and drought-resistant crops will help alleviate current and future world food demand by enabling production in marginal areas to expand. Improvements will be critical, as world population continues to increase, with or without climate change

Source: After Smith et al. (1995), table 3.

we may well be able to adapt to the most likely ways in which the climate may change. For example, we could substitute heat- and drought-resistant crops for those whose yields are reduced. Infrastructure is generally replaced on a much faster time-scale than climatic change; so it could be adapted to changes in climate. It can also be argued in favour of reactive adaptation that it does not involve prematurely spending money in advance of uncertain changes.

On the other hand, one can argue that rapid climate change, or significant increases in the intensity and frequency of extreme events such as floods, storms or droughts, could make reactive adaptations difficult and could pose immediate problems for large numbers of people. Equally, some policies would have significant benefits even under current environmental conditions and would be valuable from a cost–benefit perspective even if no climatic change took place. These types of anticipatory policies are often called 'no regrets policies' because they will succeed whether or not climatic change takes place, meaning that policy-makers should never have to regret their adoption. 'No regrets' policies may, none the less, be expensive. Table VII.2 illustrates a selection of these policies.

One can argue that the central challenge for policy-makers in coming decades will be to find ways of allowing the global economy to grow at a moderate rate while at the same time maintaining or enhancing the protection of wilderness, the prevention of pollution, and the sustenance of ecological resources. We cannot be sure that we will find policies that enable this to happen. Governments and society will inevitably need to make difficult trade-offs between economic growth and environmental protection. We cannot envisage a situation where there is indefinite growth in the human population and indefinite growth in the consumption of resources. We need to ensure, to use Sir Crispin Tickell's phrase (Tickell, 1993) that humans are not 'a suicidal success'.

KEY TERMS AND CONCEPTS

anticipatory adaptation
no-regrets policies
reactive adaptation

POINTS FOR REVIEW

Which environmental issues will become increasingly important in coming decades?

Can human energies be harnessed over the coming decades to improve rather than to degrade the environment?

How might we adapt to global warming should it occur?

GLOSSARY

In each definition, any words that themselves appear in the glossary are printed in *italic* type.

adiabatic compression The process by which, as a parcel of air falls, the internal energy is increased and its temperature is raised.

acid rain Rain which, because of the presence of dissolved substances derived from air pollution, has a *pH* of less than 5.65.

aerosol (atmospheric) An aggregation of minute particles (solid or liquid) suspended in the atmosphere. The term is often used to describe smoke, condensation nuclei, freezing nuclei or fog, or pollutants such as droplets containing sulphur dioxide or nitrogen dioxide.

aggradation The building upwards or outwards of the land surface by the deposition of sediment.

albedo A measure for the reflectivity of a body or surface, defined as the total *radiation* reflected by the body divided by the total radiation falling on it. Values are expressed on a scale of either 0–1 or 1–100%.

alluvial floodplain A flat-lying area composed of sediments (sands, silts, clays, gravels, etc.) deposited by rivers.

amphibian A creature that can live on land or in water.

anthropogenic Caused by human activities.

anthropogeomorphology The study of the human impact on landforms and landforming processes.

aquaculture The cultivation or rearing of plants or animals that grow or live in or near water.

aquifer An underground water-bearing layer of porous rock through which water can flow.

arid Dry, with limited vegetation, rainfall less than about 250 mm and a great excess of evaporation over *precipitation*.

arterial drainage A system of major drainage channels into which numerous small channels feed.

atoll An irregular, annular (ring-shaped) coral algal reef enclosing or almost enclosing a central lagoon. The reefs are often breached by channels.

backscatter To send back, rather than let through, incoming *radiation* from the sun.

badlands Areas that have been eroded by deep systems of ravines or gullies.

barrier island An elongated, mainly sandy, ridge feature running parallel to the coast and separated from it by a lagoon.

base level The lower limit down to which erosion on land may operate, usually defined with reference to the role of running water. For example, sea level acts as a general base level, though there can be a wide range of local base levels above and below sea level.

basin The area that drains into a particular river. It has the same general meaning as *catchment* (British usage) or *watershed* (American usage).

biodegradable A term used to describe a substance that can be rendered harmless or be broken down by natural processes.

biodiversity A term used to describe the variety of species, both floral and faunal, contained within an *ecosystem*.

biofabrics Fabrics made of organic material.

biological magnification The increased

concentration of toxic material at consecutive, higher *trophic levels* in an *ecosystem*. Toxins, such as *heavy metals* and persistent pesticides, become incorporated into living tissue from the environment.

biomass The total mass of biological material contained in a given area of the Earth's surface (expressed as dry weight per unit area).

biome A major ecological community or complex of communities that extends over a large geographical area and is characterized by a dominant type of vegetation (e.g. *tundra*, desert, rain forest).

bioremediation The use of micro-organisms to restore the qualities of environments contaminated by hazardous substances.

biosphere The interlinked communities of animals, plants and micro-organisms that live on the land and sea of the Earth.

biota The animal and plant life of a region.

biotechnology The manipulation of living organisms and their components (e.g. genes or gene components) for specific tasks.

bloom A scum produced by algae on the surface of standing water.

blowout An area of dune that has been breached by wind excavation.

boreal Of northern regions. A term applied both to a climatic zone characterized by cold, snowy winters and short summers, and to the coniferous forests of the high mid-latitudes in the Northern Hemisphere, also known as taiga.

brecciate Break up into angular fragments.

carbon budget The balance between the amount of carbon which accumulates in a system and the amount that is released.

carcinogen Any substance that produces cancer.

carrying capacity The maximum population of a given organism which a particular environment can sustain without a tendency to decrease or increase.

catalyst A substance that, without itself undergoing any permanent change, sets off a change or increases the rate at which a change occurs.

catchment The area that drains into a river. It is bounded by a drainage divide or *watershed* (British usage).

centre–pivot irrigation The artificial distribution of water to land for agricultural use in which *groundwater* is pumped and from a central point is dispersed in a circle.

channelization The modification of river channels for the purpose of flood control, land drainage, navigation and the reduction or prevention of erosion.

chaparral A type of stunted (*scrub*) woodland found in *temperate* regions with dry summers. It is dominated by drought-resistant evergreen shrubs.

chlorofluorocarbons A range of synthetically manufactured, chemically inert compounds containing atoms of carbon, fluorine and chlorine. They have been developed and widely used as solvents, refrigerants and aerosol propellants and in the manufacture of foam plastics.

colloidal Composed of ultramicroscopic particles.

convection The transfer of heat in the atmosphere by the upward flow of hot air or the downward flow of cold air.

deflation The removal of dry, unconsolidated material, e.g. dust or sand, from a surface by wind.

deflocculate To disperse or break up an aggregate so that particles become suspended in a solution. This may be achieved by the presence of sodium cations.

defoliant An agent that removes foliage (e.g. leaves) from a plant.

deforestation The permanent removal of trees from an area of forest or woodland.

desertification The spread of desert-like conditions in *arid* or *semi-arid* areas, due to human interference or climatic change or both.

desiccation   Drying up of the environment.

diatom   A microscopic single-celled alga with a siliceous cell wall.

dieback   A diseased condition of plants, often applied to the dying-off of large tracts of similar species at the same time.

dimethylsulphide   A volatile sulphur compound in seawater produced by bacterial decay and planktonic algae. It oxidizes in the atmosphere to form a sulphate *aerosol*.

discharge (rivers)   The amount of water that flows in a river.

DNA (deoxyribonucleic acid)   The substance that is the carrier of *genetic* information, found in the chromosomes of the nucleus of a cell.

domestication   The taming and breeding of previously wild animals and plants for human use.

drainage basin   That part of the land surface which is drained by a particular river system and is defined by a divide or *watershed* (British usage)

drawdown   The reduction in *groundwater* level by pumping out water faster than it can be replenished.

dust storm   A storm in a *semi-arid* area which carries dense clouds of dust, sometimes to a great height, often obscuring visibility to below 1,000 metres.

ecology   The science which studies the relations between living organisms and their environment.

ecosystem   A biological community of any scale in which organisms interact with their physical environment.

ecotone   A transition zone marking an overlap rather than a distinct boundary between two plant communities. It may be a zone of tension.

edaphic   A term used to describe soil conditions which influence the growth of plants and other organisms. Edaphic factors include physical, chemical and biological properties of soils, such as

pH, particle-size distribution and organic content.

El Niño events   A term applied to the extensive, intense and prolonged warming of the eastern tropical Pacific Ocean which occurs every few years. It is associated with major anomalies in the patterns of atmospheric circulation and rainfall.

endemic   Normally found only among a particular people or in a certain region.

eustasy   A worldwide change in sea level, indicating an actual rise or fall of the sea.

eutrophication   The process by which an aquatic *ecosystem* increases in productivity as a result of increased nutrient input. Often this is due to human-induced additions of elements such as nitrogen and phosphorus. However, the process may also be a natural phenomenon.

evapotranspiration   The combined loss of water by evaporation from the soil surface and transpiration from plants.

ex situ **methods**   A term used to describe means of conserving species outside their natural habitat (e.g. in zoos or botanic gardens).

feral   Term describing an animal or plant, once *domesticated*, that has gone wild.

filling   The deposition of dredged material to make new land.

fluvial   Relating to a river or rivers.

food chain   The transfer of energy from green plants through a sequence of organisms in which each eats the one below it in the chain and is eaten by the one above.

forest decline   The decline of forest vitality characterized by decreased and abnormal growth, leading eventually to death. The causes are poor management practices; climatic change; fungal, viral and pest attack; nutrient deficiency; and atmospheric pollution.

friable   Easily crumbled (of soil, rock, or other material).

gabion   A wire-framed container full of boulders or cobbles, used to make walls to stop erosion.

general circulation model (GCM) A dynamic computer model which simulates large-scale features of atmospheric and oceanic circulation.

genetic Relating to genes, which are units of heredity composed or *DNA* or *RNA* and forming part of a chromosome that determines the particular characteristics of an individual.

geomorphology The science of the origin and development of landforms.

glaciated Term used to describe an area that has been at some point covered or moulded by glaciers or ice sheets.

global warming The process by which the Earth may become warmer because of the role of mechanisms such as the *greenhouse effect*.

gneiss A coarse-grained metamorphic rock composed of feldspars, quartz and ferromagnesian minerals.

greenhouse effect A climatic effect caused by permitting incoming solar *radiation* but inhibiting outgoing radiation. Incoming short-wave radiation is absorbed by materials which then re-radiate longer wavelengths. Certain substances in the atmosphere, e.g. carbon dioxide, absorb long-wave radiation, resulting in a warming effect.

gross primary production The total amount of organic material synthesized in a given time period by living organisms from inorganic material.

groundwater Water occurring below the soil surface that is held in the soil itself or in a deeper *aquifer*.

gypsum A rock formed of natural calcium sulphate caused by its crystallization as salty water is concentrated by evaporation.

habitat The place in which an organism lives, characterized by its physical features or the dominant plant types.

halons Members of the halogenated fluorocarbon (HF) group of ethane- or methane-based compounds in which H$^+$ ions are partially or completely replaced by chloride, fluoride and/or bromide. They are long-lived and have been implicated in ozone depletion.

halophytic Tolerant of high concentrations of salts.

heathland An area of evergreen *sclerophyllous* shrubland where heath families (e.g. Ericaceae) are present, though not necessarily dominant. Heathlands develop on areas where soil is low in nutrient status.

heavy metal Any metal or alloy of high specific gravity, especially one that has a density higher than 5 g per cu cm, e.g. lead, zinc, copper, mercury.

herbicide Any agent, organic or inorganic, used to destroy unwanted vegetation.

Holocene The most recent epoch of the Quaternary, following the Pleistocene. Often called the post-glacial, it has extended from about 10,000 years ago to the present day. It has been marked by various climatic fluctuations.

humus The organic constituent of a soil, usually formed by the decomposition of plants and leaves.

hydro-isostasy The reaction of the Earth's crust to the application and removal of a mass of water. For example, *eustatic* sea-level changes have affected the depth of water over the continental shelves, causing the crust to be depressed at times of high sea level and elevated at times of low sea level.

hydrocarbons Compounds of hydrogen and carbon, some with minor or trace quantities of oxygen, sulphur, nitrogen and other elements.

hydrocompaction The process by which sediments are compressed by an overlying body of water.

hydrology The science concerned with the study of the different forms of water as they exist in the natural environment. Its central focus is the circulation and distribution of water.

hydrostatic uplift Uplift of land surface caused by upward water pressure.

hypoxia   The condition experienced when oxygen levels are low in blood and tissues.

inbreeding   Breeding from closely related animals or persons.

infiltration   The movement of water into the soil from the ground surface.

interglacial   A time period between two glacial stages, during which temperatures are relatively high.

interpluvial   A time period between two *pluvial* stages, during which conditions are relatively dry.

isopleth   Line on map connecting places where a particular meteorological factor, e.g. thunderstorms, occurs with the same frequency.

isostasy   A process that causes the Earth's crust to rise or sink according to whether a weight is removed or added to it. Such a weight could be, for example, an ice cap.

karst   A limestone region with underground drainage and many cavities and passages caused by the solution of the rock.

landfill   The disposal of waste by tipping it on land, often in old mine workings or low-lying land.

laterite   The residual deposits formed by the chemical weathering of rock, composed primarily of hydrated iron and aluminium oxides. Extensively developed in the humid or subtropical regions.

leachate   The solution or soluble material that results from a *leaching* process.

leaching   The removal of dissolved material by the percolation of water through a soil or sediment.

Lessepsian migration   An almost unidirectional migration of *biota* from one sea or lake to another. Named after the man who built the Suez Canal, which allowed organisms to pass from the Red Sea to the Mediterranean.

levée   A natural or man-made embankment along a river.

lichenometry   A method of time estimation (dating) on rock surfaces based on the rate of growth of lichens (e.g. *Rhizocarpon geographicum*).

lithosphere   The solid earth.

loess   A deposit of primarily silt-sized material that was originally dust transported by the wind.

macrobenthic   Relating to large organisms that live on or near the bottom of a body of water.

macropore   A particularly large pore or void in the soil.

mammal   A warm-blooded creature with a backbone which, if female, can nourish its young.

mangrove   Plant communities dominated by mangrove trees, *Rhizophera*, *Bruguieria* and *Avicennia*, which colonize tidal mudflats, estuaries and other sheltered areas in tropical and subtropical areas.

maquis   *Scrub* vegetation of evergreen shrubs, characteristic of the western Mediterranean; broadly equivalent to *chaparral*.

marginal land   Land that is difficult to cultivate or unprofitable.

mariculture   Farming of the sea.

marsupial   A *mammal* characterized by being born incompletely developed and so usually carried and suckled in a pouch on the mother's belly for a time.

meander   The winding pattern of a sinuous river channel.

Mediterranean climate   A climatic type characteristic of the western margins of continents in the world's warm *temperate* zones between latitudes 30° and 40° (e.g. central Chile, central California).

megafauna   The largest types of animals in a community.

Mesolithic   A cultural period following the Palaeolithic, from 10,000 BC to 4,000 BC, characterized by the use of microlithic implements.

metamorphic   Term used to describe rocks which have been altered by external sources of heat, pressure or chemical substances rather than merely by burial under other rock.

metapedogenesis Human modification of soils.

microclimate The physical state of the atmosphere close to a very small area of the Earth's surface, often in relation to living matter such as crops or insects.

monsoon A wind with seasonal reversals of direction.

morphology The form or shape of an object or organism.

nanoplankton The smallest of the *phytoplankton*.

necrosis The localized death of cells, tissue or an organ resulting from disease or injury.

Neolithic A cultural period following the *Mesolithic*, from the fourth millennium BC until the onset of the Bronze Age. It marks the beginning of the *domestication* of animals and the cultivation of crops.

net biological primary production See *net primary production*.

net primary production The amount of organic material produced by living organisms from inorganic sources in excess of that used in respiration.

nutrient sink A location in which nutrients accumulate.

nutrient source A location from which nutrients are released.

oceanic conditions Climatic conditions that are modified by the presence of a nearby sea or ocean, in contrast to continental conditions.

omnivore An animal which eats both plant and animal matter.

organochlorides Organic compounds which contain chlorine. Often used as active ingredients for pesticides, they are very persistent, due to their chemical stability and low solubility. An example is the insecticide DDT.

orographic A term used to describe climatic conditions or phenomena caused by the presence of high relief (e.g. mountains).

osmosis The passage of a solvent through a semi-permeable partition or membrane into a more concentrated solution.

oxidation A chemical reaction in which a substance decreases its number of electrons. The most frequent oxidant is molecular oxygen.

palaeolimnology The study of the environmental history of a lake, most importantly from evidence preserved in its bottom sediments.

pastoralism A form of land use relating to flocks and herds of animals.

pathogen An organism which causes disease.

per capita For each person.

perennial Lasting through a year or several years. Used to describe plants that are not merely annuals and streams that normally flow through all seasons of the year.

permafrost The thermal conditions in soil and rock where temperatures are below 0°C for at least two consecutive years.

pH The measure of the acidity or alkalinity of a substance based on the number of hydrogen ions present in a litre of the substance and expressed in terms of $pH = \log_{10}(1/H^+)$ where H^+ is the hydrogen ion concentration. The centre point on the scale is 7, representing neutrality. Acid substances have a pH of less than 7 and alkaline substances have a pH of more than 7.

photochemical reaction A chemical reaction which is speeded up by particular wavelengths of electromagnetic *radiation*.

phytoplankton Microscopic organisms, especially algae, that live near the surface of the sea and form the basis of food for many other forms of aquatic life.

piezometric surface A subterranean surface marking the level to which water will rise within an *aquifer*.

Pleistocene The first epoch of the Quaternary, including glacial and *interglacial* stages, between about 2 million and 10,000 years ago.

pluvial A climatic phase with plentiful moisture.

podzol A soil characterized by the acidification of the A horizon, the downward *leaching* of cations, metals and humic substances and their deposition in the B horizon, often precipitating to form a pan. The process is most prominent in cool and wet climates.

pollen analysis The analysis of plant pollen under the microscope to reconstruct the vegetation conditions under which the sediment in which it occurs was deposited.

precipitate In chemistry, the deposition in solid form from a solution.

precipitation Moisture that falls on the ground, including rain, snow, dew and fog.

predator An animal which kills others for food by preying on them. A secondary consumer in a *food chain*.

profile An outline seen from one side (e.g. the cross profile of a river channel) or a vertical cross-section (e.g. of a soil and its various layers).

radiation, solar Electromagnetic waves emitted by the sun.

radiation budget A term used in meteorology to describe the difference between incoming and outgoing *radiation*.

radiocarbon dating A method of determining the age of an organic material (e.g. wood, charcoal, peat) by measuring the proportion of the ^{14}C isotope contained within its carbon content.

rangeland A large area of open land used for grazing or hunting.

reclamation Bringing land into a new form. This can involve either returning something to its original state (e.g. some degraded land) or transforming it into a new state (e.g. by filling in a lake to make land).

redox potential A measurement of the willingness of an electron carrier to act as a reducing or oxidizing agent.

rendzina A type of soil, with dark surface layers or horizons, that developes on soft limestones.

rill A small channel in a soil or rock surface, often only a few centimetres long.

riparian Of or on a river bank.

rip-rap Large fragments of broken rock dumped along a shoreline to protect it against wave action.

runoff The water leaving a drainage area. It is normally regarded as the rainfall minus the loss by evaporation.

salinization The process whereby salts, e.g. sulphates, nitrates and chlorides, become concentrated in the soil.

sanitization The process by which something is made more sanitary, hygienic or disinfected so that health conditions are improved.

saturation excess overland flow Surface *runoff* that is generated when rain falls on ground that is already saturated with water.

savanna A grassland of the tropics and subtropics.

scarification The process by which seeds are cleaned by abrasion of the epidermis. Can also refer to changes caused to seeds by passing through the gut of an animal or by fire.

schist A *metamorphic* rock composed of layers of different materials split into thin, irregular plates.

sclerophyllous A term referring to species of evergreen trees and shrubs that have adapted to lengthy seasonal drought.

scrub A type of vegetation consisting mainly of brushwood or stunted forest growth.

secondary forest Woodland which has regenerated and colonized an area after the original forest has been removed.

sediment yield, sediment load Sediment yield is the mean sediment load carried by a stream, giving some measure of the rate of erosion in a *drainage basin*. The sediment yield is expressed as weight per unit area.

seedbed An area of soil in which seeds are planted and take root.

seep An area moistened by the seepage of water from or into the ground.

semi-arid Dry, with a shortage of moisture for much of the year, but not so dry as an *arid* area.

shear strength The maximum resistance of a material to the application of stress. Major sources of such resistance are cohesion and friction.

sheet flow The flow of water in thin films over a low-angle surface.

shifting cultivation Cultivation of a small area of land in which forest is cleared and the *biomass* removed or burned, followed by the use of the site for the production of mixed agricultural crops for several years. Eventually the area is abandoned as soil fertility decreases, and the cultivators move on to a new area. *Slash-and-burn* is a type of shifting cultivation.

sink-hole A hole or depression in the landscape into which water drains, caused by concentration of solution of the bedrock, usually limestone or chalk.

slash-and-burn A system of land use, especially prevalent in the tropics, in which land is cleared of forest by cutting and burning so that cultivation can take place. As fertility rapidly declines in the cultivated areas, the farmer moves on to a new area after a few years.

smectite A type of clay, often made up of montmorillonite, that may have the property of swelling in water.

smog A fog in which smoke or other forms of atmospheric pollutants play an important role in causing the fog to form and thicken. It often has unpleasant or dangerous physiological effects.

splash erosion Erosion produced by the impact of raindrops splashing on the ground surface, particularly if it is not protected by vegetation.

spontaneous combustion Fire caused by the natural build-up of heat within inflammable material.

spp. Abbreviation for 'species' (plural).

steppe A generally dry, grassy plainland.

stratosphere The region of the atmosphere lying between the *tropopause* and about 20 km, in which there is little change in temperature with height.

substrates Material underlying the surface.

succession The sequence of changes in a plant community as it develops over time.

supernatant Term describing liquid floating on a surface.

sustainable development Development that meets the needs of the present without compromising the ability of future generations to meet their own needs.

symbiosis An interaction between two different organisms living in close contact and usually to the advantage of both.

talus A sloping mass of fragments, similar to scree, at the foot of a cliff.

tectonic A term describing the broad structures of the Earth's *lithosphere* and movements within the Earth's crust.

temperate A term used to describe a region or climate characterized by mild temperatures.

temperature inversion Normally, air temperature decreases as height increases. However, under certain weather conditions air temperature may increase with height so that a layer of warmer air overlies a colder layer. This is temperature inversion.

terracing The construction of banks or steps on a hillside to give areas of low gradient, either to enable cultivation or to conserve soil.

thalweg Line where opposite slopes meet at the bottom of a valley, river or lake.

thermokarst Topographical depressions resulting from the thawing of ground ice.

threshold A condition which marks the transition from one state of operation of a system to another. Rapid and irreversible change may occur.

trace elements Elements that are required by living organisms to ensure normal growth, development and maintenance.

They occur at lower concentrations than major elements and include iron, manganese, zinc, copper, iodine, etc.

trace gases   Gases which occur in very small amounts in the atmosphere.

trophic levels   The positions that organisms occupy in a *food chain*.

tropopause   The interface between the *troposphere* and the *stratosphere*.

troposphere   The lowest level of the atmosphere, in which most of our weather occurs. It lies beneath the stratosphere and its thickness ranges from about 7 km at the poles to about 28 km at the equator.

tundra   The zone between the latitudinal limits of tree growth and polar ice, characterized by severe winters and a short growing season.

turbidity   A measure of the lack of clearness in a liquid caused by the presence of suspended material.

understory   A layer of vegetation beneath the main tree canopy.

UV radiation   Radiation from the sun with shorter wavelengths than visible light. It is classified into three ranges according to its effect on human skin: UV-A is not normally harmful; UV-B produces reddening and tanning; UV-C (with the shortest wavelengths) is the most damaging.

vector-borne   A term used to describe a disease that is passed on by an organism, often an insect (e.g. as malaria is transmitted by the mosquito).

volatilization   Evaporation, or the process of turning from solid or liquid form into a vapour.

water table   The level below which the ground is saturated with water.

watershed (American usage)   The area occupied by a *drainage basin* or stream *catchment*.

watershed (British usage)   A line of separation between waters flowing into different rivers, *basins* or seas.

weather front   A sloping boundary surface separating two air masses that exhibit different meteorological properties.

wetlands   The collective term for *ecosystems* whose formation has been dominated by water, and whose processes and characteristics are largely controlled by water.

wilderness   An area left untouched and thus in a natural state, with little or no human control or interference.

wind reactivation   The renewed movement of sand and other material by the wind, especially when vegetation cover is reduced.

wind throw   The blowing over of trees by the wind.

xerophilous   A term describing plants which live in dry *habitats* and can endure prolonged drought. Many such plants, e.g. cactus, have developed physiological adaptations to cope with these conditions.

REFERENCES

Abu-Atta, A. A., 1978, *Egypt and the Nile after the Construction of the High Aswan Dam*. Cairo: Ministry of Irrigation and Land Reclamation.

Adams, W. M., 1993, Indigenous use of wetlands and sustainable development in West Africa. *Geographical Journal* 159, 209–18.

Aiken, S. R. and Leigh, C. H., 1992, *Vanishing Rainforests: Their Ecological Transition in Malaysia*. Oxford: Oxford University Press.

Al-Ibrahim, A. A., 1991, Excessive use of ground-water resources in Saudi Arabia: impacts and policy options. *Ambio* 20, 34–7.

Alpert, P., 1993, Conserving biodiversity in Cameroon. *Ambio* 22, 44–8.

Anderson, D. M., 1994, Red tides. *Scientific American* 271(2), 52–8.

Andreae, M. O., 1991, Biomass burning: its history, use and distribution and its impact on environmental quality and global climate. In J. S. Levine (ed.), *Global Biomass Burning*, 3–21. Cambridge, Mass.: MIT Press.

Atkinson, B. W., 1968, A preliminary examination of the possible effect of London's urban area on the distribution of thunder rainfall, 1951–60. *Transactions, Institute of British Geographers* 44, 97–118.

Attewell, P., 1993, *Ground Pollution*. London: Spon.

Bakan, S., Chlono, A., Cubasch, U., Feichter, J., Graf, H., Grassl, H., Hasselman, K., Kirchner, I., Latif, M., Roeckner, E., Samsen, R., Schlese, U., Schrivener, D., Schult, I., Sielman, F. and Wells, W., 1991, Climate response to smoke from the burning oil wells in Kuwait. *Nature* 351, 367–71.

Bari, M. A. and Schofield, N. J., 1992, Lowering of a shallow, saline water table by extensive eucalypt reforestation. *Journal of Hydrology*, 133, 273–91.

Beveridge, M. C. M., Ross, L. G. and Kelly, L. A., 1994, Aquaculture and biodiversity. *Ambio* 23, 497–503.

Bidwell, O. W. and Hole, F. D., 1965, Man as a factor of soil formation. *Soil Science* 99, 65–72.

Bird, E. C. F., 1985, *Coastline Changes*. Chichester: Wiley.

Blackburn, W. H., Knight, R. W. and Schuster, J. L., 1983, Saltcedar influence on sedimentation in the Brazos River. *Journal of Soil and Water Conservation* 37, 298–301.

Boardman, J., 1992, Current erosion on the South Downs: implications for the past. In M. Bell and J. Boardman (eds), *Past and Present Soil Erosion*, 9–19. Oxford: Oxbow Books.

Boardman, J., 1995, Damage to property by runoff from agricultural land, South Downs, southern England, 1976–1993. *Geographical Journal* 161, 177–91.

Boardman, J., Foster, I. D. L. and Dearing, J. A. (eds), 1990, *Soil Erosion on Agricultural Land*. Chichester: Wiley.

Boehmer-Christiansen, S. and Skea, J., 1991, *Acid Politics: Environment and Energy Policies in Britain and Germany*. London: Belhaven Press.

Boorman, L. A., 1993, Dry coastal ecosystems of Britain: dunes and shingle beaches. In E. van der Maarel (ed.), *Dry Coastal Ecosystems*, 197–228. Amsterdam: Elsevier.

Bridgman, H., Warner, H. and Dodson, J., 1995, *Urban Biological Environments*. Melbourne: Oxford University Press.

Brimblecombe, P., 1977, London air pollution 1500–1900. *Atmospheric Environment* 11, 1157–62.

Brimblecombe, P., 1987, *The Big Smoke*. London: Methuen.

Brookes, A., 1985, River channelization: traditional engineering methods, physical consequences, and alternative practices. *Progress in Physical Geography* 9, 44–73.

Brookes, A., 1987, The distribution and management of channelized streams in Denmark. *Regulated Rivers* 1, 3–16.

Brookes, A., 1988, *Channelized Rivers*. Chichester: Wiley.

Browning, K. A., Allah, R. J., Ballard, B. P., Barnes, R. T. H., Bennetts, D. A., Maryon, R. H., Mason, P. J., McKenna, D., Mitchell, J. F. B., Senior, C. A., Slingo, A. and Smith, F. B., 1991, Environmental effects from burning oil wells in Kuwait. *Nature* 351, 363–7.

Bryson, R. A. and Barreis, D. A., 1967, Possibility of major climatic modifications and their implications: northwest India, a case for study. *Bulletin of the American Meteorological Society* 48, 136–42.

Budyko, M. I., 1974, *Climate and Life*. New York: Academic Press.

Calder, I., 1992, Hydrologic effects of land-use change. In D. R. Maidment (ed.), *Handbook of Hydrology*, 13.1–13.50. New York: McGraw-Hill.

Carrera, F., 1993, Computerised catalog of outdoor art in Venice with automatic estimation of restoration costs. In M.-J. Thiel (ed.), *Conservation of Stone and Other Materials*, 831–8. London: Spon.

Carter, F. W. and Turnock, D. (eds), 1993, *Environmental Problems in Eastern Europe*. London: Routledge.

Chandler, T. J., 1976, The climate of towns. In T. J. Chandler and S. Gregory (eds), *The Climate of the British Isles*, 307–29. London: Longman.

Charlson, R. J., Schwartz, S. E., Hales, J. M., Cess, R. D., Coakley, J. A., Hansen, J. E. and Hoffmann, D. J., 1992, Climate forcing by anthropogenic aerosols. *Science* 255, 423–30.

Charney, J., Stone, P. H. and Quirk, W. J., 1975, Drought in the Sahara: a biogeophysical feedback mechanism. *Science* 187, 434–5.

Clark, J. A., Farrell, W. E. and Peltier, W. R., 1978, Global changes in postglacial sea level: a numerical calculation. *Quaternary Research* 9, 265–87.

Clark, R. B., 1989, *Marine Pollution*, 2nd edn. Oxford: Clarendon Press.

Cooke, R. U. and Doornkamp, J. C., 1993, *Geomorphology in Environmental Management*, 2nd edn. Oxford: University Press.

Cooke, R. U. and Gibbs, G., 1994, *Crumbling Heritage: Studies of Stone Weathering in Polluted Atmospheres*. Report for National Power plc.

Corlett, R. T., 1995, Tropical secondary forests. *Progress in Physical Geography* 19, 159–72.

Corrie, I. D., and Werner, P. A., 1993, Alien plant species invasive in Kakadu National Park, tropical Northern Australia. *Biological Conservation* 63, 127–35.

Costa, J. E. and Baker, V. R., 1981, *Surficial Geology: Building with the Earth*. New York: Wiley.

Crutzen, P. J. and Goldammer, J. G., 1993, *Fire in the Environment*. Chichester: Wiley.

Del Monte, M. and Vittori, O., 1985, Air pollution and stone decay: the case of Venice. *Endeavour* 9, 117–22.

Di Castri, F., 1989, History of biological invasions with special emphasis on the old world. In W. C. Clark and R. E. Munn (eds), *Sustainable Development of the Biosphere*, 252–89. Cambridge: Cambridge University Press.

Dikau, R., Brunsden, D., Schrott, L. and Ibsen, M-L., 1996, *Landslide Recognition*. Chichester: Wiley.

Dobson, M., 1991, De-icing salt damage to trees and shrubs. *Forestry Commission Bulletin*, no. 101.

Douglas, T., 1992, Patterns of land, water and air pollution by wastes. In M. Newson (ed.), *Managing the Human Impact on the Natural Environment: Patterns and Processes*, 150–71. London: Belhaven Press.

Downing, R. A. and Wilkinson, W. B. (eds), 1991, *Applied Groundwater Hydrology: A British Perspective*. Oxford: Clarendon Press.

Drake, J. A. (ed.), 1989, *Biological Invasions: A Global Perspective*. Chichester: Wiley.

Dunne, T. and Leopold, L. B., 1978, *Water in Environmental Planning*. San Francisco: Freeman.

Edmonds, R. L., 1994, *Patterns of China's Lost Harmony: A Survey of the Country's Environmental Degradation and Protection*. London: Routledge.

Ehrlich, P. R. and Ehrlich, A. H., 1982. *Extinction*. London: Gollancz.

Ellenberg, H., 1979, Man's influence on tropical mountain ecosystems in South America. *Journal of Ecology* 67, 401–16.

Elsom, D., 1992, *Atmospheric Pollution*, 2nd edn. Oxford/Cambridge, Mass.: Blackwell.

Elsom, D., 1996, *Smog Alert*. London: Earthscan.

Elton, C. S., 1958, *The Ecology of Invasions by Plants and Animals*. London: Methuen.

Englefield, G. J. H., Tooley, M. J. and Zhang, Y., 1990, *An Assessment of the Clwyd Coastal Lowlands after the Floods of February 1990*. Environmental Research Centre, University of Durham, Publication no. 4.1.

Fillenham, L. F., 1963, Holme Fen Post. *Geographical Journal* 129, 502–3.

Freedman, B., 1995, *Environmental Ecology*, 2nd edn. San Diego: Academic Press.

Fullen, M. A. and Mitchell, D. J., 1994, Desertification and reclamation in North Central China. *Ambio* 23, 131–5.

GESAMP, 1990, *The State of the Marine Environment*. Oxford: Blackwell Scientific.

Giddings, J., 1973, *Chemistry, Man and Environmental Change*. San Francisco: Canfield Press.

Gimingham, C. H. and de Schmidt, I. T., 1983, Heaths and natural and semi-natural vegetation. In W. Holzner, M. J. A. Werger and I. Ikusima (eds), *Man's Impact on Vegetation* 185–99. The Hague: Junk.

Gleick, P. H. (ed.), 1993, *Water in Crisis: A Guide to the World's Freshwater Resources*. New York: Oxford University Press.

Gomez, B. and Smith, C. G., 1984, Atmospheric pollution and fog frequency in Oxford, 1926–80. *Weather* 39, 379–84.

Goudie, A. S., 1990, *The Landforms of England and Wales*. Oxford: Blackwell.

Goudie, A. S., 1993, *The Human Impact on the Natural Environment*, 4th edn. Oxford: Blackwell.

Goudie, A. S., 1995, *The Changing Earth: Rates of Geomorphological Processes*. Oxford: Blackwell.

Goudie, A. S. (ed.) 1985, *Encyclopaedic Dictionary of Physical Geography*. Oxford: Blackwell.

Goudie, A. S. (ed.), 1990, *Techniques for Desert Reclamation*. Chichester: Wiley.

Goudie, A. S. and Middleton, N. J., 1992, The changing frequency of dust storms through time. *Climatic Change* 20, 197–225.

Gowlett, J. A. J., Harris, J. W. K., Walton, D. and Wood, B. A., 1981, Early archaeological sites, hominid remains and traces of fire from Chesowanja, Kenya. *Nature* 284, 125–9.

Graetz, D., 1994, Grasslands. In W. B. Meyer and B. L. Turner (eds), *Changes in Land Use and Land Cover: A Global*

Perspective, 125–47. Cambridge: Cambridge University Press.

Graf, W. L., 1985, *The Colorado River: Instability and Basin Management*. Washington DC: Association of American Geographers.

Grainger, A., 1990, *The Threatening Desert: Controlling Desertification*. London: Earthscan.

Grainger, A., 1992, *Controlling Tropical Deforestation*. London: Earthscan.

Green, F. H. W., 1978, Field drainage in Europe. *Geographical Journal* 144, 171–4.

Gregory, K. J., 1985, The impact of river channelization. *Geographical Journal* 151, 53–74.

Gribbin, J., 1988, *The Hole in the Sky: Man's Threat to the Ozone Layer*. London: Corgi Books.

Guilcher, A., 1988, *Coral Reef Geomorphology*. Chichester: Wiley.

Hammerton, D., 1994, Domestic and industrial pollution. In P. S. Maitland, P. J. Boon and D. S. McLusky (eds), *The Freshwaters of Scotland: A National Resource of International Significance*, 247–64. Chichester: Wiley.

Hardoy, J. E., Mitlin, D. and Satterthwaite, D., 1992, *Environmental Problems in Third World Cities*. London: Earthscan.

Harris, D. R. (ed.), 1980, *Human Ecology in Savanna Environments*. London: Academic Press.

Hawkins, J. P. and Roberts, C. M., 1994, The growth of coastal tourism in the Red Sea: present and future effects on coral reefs. *Ambio* 23, 515–18.

Helldén, U., 1984, Land degradation and land productivity monitoring: needs for an integrated approach. In A. Hjört (ed.), *Land Management and Survival*, 77–87. Uppsala: Scandinavian Institute of African Studies.

Hoffman, M., 1991, Taking stock of Saddam's fiery legacy in Kuwait. *Science* 253, 971.

Hollis, G. E., 1978, The falling levels of the Caspian and Aral Seas. *Geographical Journal* 144, 62–80.

Houghton, J. T., 1994, *Global Warming: The Complete Briefing*. Oxford: Lion.

Houghton, J. T., Callander, B. A. and Varney, S. K. (eds), 1992, *Climate Change 1992: The Supplementary Report of the IPCC Scientific Assessment*. Cambridge: Cambridge University Press.

Houghton, J. T., Jenkins, G. J. and Ephraums, J. J. (eds), 1990, *Climate Change: The IPCC Scientific Assessment*. Cambridge: Cambridge University Press.

Houghton, J. T., Meira Filho, L. G., Callandar, B. A., Harris, N., Kaltenberg, A. and Maskell, K. (eds), 1996, *Climate Change 1995: The Science of Climate Change*. Cambridge: Cambridge University Press.

Hudson, N., 1971, *Soil Conservation*. London: Batsford.

Hull, S. K. and Gibbs, J. N., 1991, Ash dieback: a survey of non-woodland trees. *Forestry Commission Bulletin* 93.

Husain, T. and Amin, M. B., 1994, Kuwaiti oil fires: particulate monitoring. *Atmospheric Environment* 28, 2235–48.

Ibe, A. C., 1988, Nigeria. In H. J. Walker (ed.), *Artificial structures on shorelines*, 287–94. Dordrecht: Kluwer Academic.

Iltis, H. H., 1988, Serendipity in the exploration of biodiversity: what good are weedy tomatoes? In E. O. Wilson (ed.), *Biodiversity*, 98–105. Washington DC: National Academy Press.

Innes, J. L., 1983, Lichenometric dating of debris-flow deposits in the Scottish highlands. *Earth Surface Processes and Landforms* 8, 579–88.

Innes, J. L., 1992, Forest decline. *Progress in Physical Geography* 16, 1–64.

Innes, J. L. and Boswell, R. C., 1990, Monitoring of forest condition in Great Britain 1989. *Forestry Commission Bulletin* 94, 57.

Ives, J. D. and Messerli, B., 1989, *The*

Himalayan Dilemma: Reconciling Development and Conservation. London: Faber.

Johnson, A. T. (ed.), *Land Subsidence*. IAHS Publication no. 200.

Johnson, D. L. and Lewis, L. A., 1995, *Land Degradation: Creation and Destruction*. Oxford: Blackwell.

Jones, D. K. C. (ed.), 1993, Earth surface resources management in a warmer Britain. *Geographical Journal* 159, 124–208.

Kates, R. W., Turner, B. L. and Clark, W. C., 1990, The great transformation. In B. L. Turner, W. C. Clark, R. W. Kates, J. F. Richards, J. T. Matthews and W. B. Meyer (eds), *The Earth as Transformed by Human Action*, 1–17. Cambridge: Cambridge University Press.

Kelletat, D., 1989, Biosphere and man as agents in coastal geomorphology and ecology. *Geoökodynamik* 10, 215–52.

Kemp, D. D., 1994, *Global Environmental Issues: A Climatological Approach*, 2nd edn. London: Routledge.

Kibler, D. F. (ed.), 1982, *Urban Stormwater Hydrology*. Washington DC: American Geophysical Union.

Kingdon, J., 1990, *Island Africa: The Evolution of Africa's Rare Animals and Plants*. London: Collins.

Kirkpatrick, J., 1994, *A Continent Transformed*. Melbourne: Oxford University Press.

Kotlyakov, V. M., 1991, The Aral Sea basin: a critical environmental zone. *Moscow Environment* 33(1), 4–9, 36–8.

Kozlowski, T. T. and Ahlgren, C. C. (eds), 1974, *Fire and Ecosystems*. New York: Academic Press.

Kuntesal, G. and Chang, T. Y., 1987, Trends and relationships of O_3, NOx, and HC in the South Coast Air Basin of California. *Journal of the Air Pollution Control Association* 37, 1158–63.

La Roe, E. T., 1977, Dredging: ecological impacts. In J. R. Clarke (ed.), *Coastal Ecosystem Management*, 610–14. New York: Wiley.

Lamprey, H., 1975, The integrated project on arid lands. *Nature and Resources* 14, 2–11.

Landsberg, H. E., 1981, *The Urban Climate*. New York: Academic Press.

Langford, T. E. L., 1990, *Ecological Effects of Thermal Discharges*. London: Elsevier Applied Science.

Lanly, J. P., Singh, K. D. and Janz, K., 1991, FAO's 1990 reassessment of tropical forest cover. *Nature and Resources* 27, 21–6.

Lean, J. and Warrilow, D. A., 1989, Simulation of the regional climatic impact of Amazonian deforestation. *Nature* 342, 126–33.

Lee, D., 1994, Regional variations in long-term visibility trends in the UK, 1962–1990. *Geography* 79, 108–21.

Lently, A. D., 1994, Agriculture and wildlife: ecological implications of subsurface irrigation drainage. *Journal of Arid Environments* 28, 85–94.

Lents, J. M. and Kelly, W. J., 1993, Clearing the air in Los Angeles. *Scientific American*, October, 18–25.

Lerner, D., 1990, *Groundwater Recharge in Urban Areas*, 59–65. IAHS Publication no. 198.

Levine, J. S. (ed.), 1991, *Global Biomass Burning*. Cambridge, Mass.: MIT Press.

Littler, M. M. and Littler, D. S., 1995, Impact of CLOD pathogen on Pacific coral reefs. *Science* 267, 1256–1360.

Lugo, A. E., 1988, Estimating reductions in the diversity of tropical forest species. In E. O. Wilson (ed.), *Biodiversity*, 58–70. Washington DC: National Academy Press.

Mabbutt, J. A., 1985, Desertification of the world's rangelands. *Desertification Control Bulletin* 12, 1–11.

McCloskey, M. and Spalding, H., 1989, A reconnaissance-level inventory of the amount of wilderness remaining in the world. *Ambio* 18, 221–7.

McLean, R. F. and Woodroffe, C. D., 1994, Coral atolls. In R. W. G. Carter and C. D. Woodroffe (eds), *Coastal Evolution*, 267–302. Cambridge: Cambridge University Press.

McLusky, D. S., 1994, Tidal fresh waters. In P. S. Maitland, P. J. Boon and D. S. McLusky (eds), *The Freshwaters of Scotland: A National Resource of International Significance*, 51–64. Chichester: Wiley.

Maltby, E., 1986, *Waterlogged Wealth: Why Waste the World's Wet Places?* London: Earthscan.

Mannion, A. M., 1991, *Global Environmental Change*. Harlow: Longman.

Mannion, A. M., 1992, Acidification and eutrophication. In A. M. Mannion and S. E. Bowlby (eds), *Environmental Issues in the 1990s*, 177–95. Chichester: Wiley.

Mannion, A. M., 1995, *Agriculture and Environmental Change*. London: Wiley.

Marsh, G. P., 1864, *Man and Nature*. New York: Scribner. Ed. D. Lowenthal, 1965, Cambridge, Mass., Belknap/Harvard University Press.

Martin, P. S. and Klein, R. G., 1984, *Pleistocene Extinctions*. Tucson: University of Arizona Press.

Meadows, M. E. and Linder, H. P., 1993, A palaeoecological perspective on the origin of Afromontane grasslands. *Journal of Biogeography* 20, 345–55.

Mee, L. D., 1992, The Black Sea in crisis: a need for concerted international action. *Ambio* 21, 278–86.

Meyer, W. B., 1996, *Human Impact on the Earth*. Cambridge: Cambridge University Press.

Micklin, P. P., 1988, Desiccation of the Aral Sea: a water management disaster in the Soviet Union. *Science* 241, 1170–5.

Micklin, P. P., 1992, The Aral crisis: introduction to the special issue. *Post-Soviet Geography* 33(5), 269–82.

Middleton, N. J., 1991, *Desertification*. Oxford: Oxford University Press.

Middleton, N. J., 1995, *The Global Casino*. London: Edward Arnold.

Mintzer, I. M. and Miller, A. S., 1992, Stratospheric ozone depletion: can we save the sky? In *Green Globe Yearbook 1992*, 83–91. Oxford: Oxford University Press.

Mitsch, W. J. and Gosselink, J. G., 1986, *Wetlands*. New York: Van Nostrand Reinhold.

Morgan, R. P. C., 1995, *Soil Erosion and Conservation*. Harlow: Longman.

Moore, D. M., 1983, Human impact on island vegetation. In W. Holzner, M. J. A. Werger and I. Ikusima (eds), *Man's Impact on Vegetation* 237–48. The Hague: Junk.

Musk, L. F., 1991, The fog hazard. In A. H. Perry and L. Symons (eds), *Highway Meteorology*, 91–130. London: Spon.

Myers, N., 1979, *The Sinking Ark: A New Look at the Problem of Disappearing Species*. Oxford: Pergamon.

Myers, N., 1990, The biodiversity challenge: expanded hot spots analysis. *The Environmentalist* 10, 243–56.

Myers, N., 1992, Future operational monitoring of tropical forests: an alert strategy. In J. P. Mallingreau, R. da Cunha and C. Justice (eds), *Proceedings of the World Forest Watch Conference*, 9–14. San Jose dos Campos, Brazil.

Mylne, M. F. and Rowntree, P. R., 1992, Modelling the effects of albedo change associated with tropical deforestation. *Climatic Change* 21, 317–43.

Nash, L., 1993, Water quality and health. In P. H. Gleick (ed.), *Water in Crisis: A Guide to the World's Freshwater Resources*, 25–39. New York: Oxford University Press.

Newson, M., 1992, Patterns of freshwater pollution. In M. Newson (ed.), *Managing the Human Impact on the Natural*

Environment: Patterns and Processes, 130–49. London: Belhaven.

Newsom, M. (ed.), 1992, *Managing the Human Impact on the Natural Environment.* London: Belhaven.

Nicholson, S. E., 1988, Land surface–atmosphere interaction: physical processes and surface changes and their impact. *Progress in Physical Geography* 12, 36–65.

Nordstrom, K. F., 1994, Developed coasts. In R. W. G. Carter and C. D. Woodroffe (eds), *Coastal Evolution,* 477–509. Cambridge: Cambridge University Press.

Nriagu, J. O. and Pacyna, J. M., 1988, Quantitative assessment of worldwide contamination of air, water and soils by trace metals. *Nature* 337, 134–9.

OECD, 1986, *Control of Water Pollution from Urban Runoff.* Paris: Organisation for Economic Cooperation and Development.

Oke, T. J., 1987, *Boundary Layer Climates,* 2nd edn. London: Routledge.

Otterman, J., 1974, Baring high albedo soils by overgrazing: a hypothesised desertification mechanism. *Science* 186, 531–3.

Papathanassiou, E. and Zenelos, A., 1993, A case of recovery in benthic communities following a reduction in chemical pollution in a Mediterranean ecosystem. *Marine Environmental Research* 36, 131–52.

Park, C. C., 1987, *Acid Rain: Rhetoric and Reality.* London: Methuen.

Park, C. C., 1992, *Tropical Rainforests.* London: Routledge.

Peck, A. J., 1978, Salinization of non-irrigated soils and associated streams: a review. *Australian Journal of Soil Research* 16, 157–68.

Peierls, B. L., Caraco, N. F., Pace, M. L. and Cole, J. J., 1991, Human influence on river nitrogen. *Nature* 350, 386.

Perry, A. H., 1981, *Environmental Hazards in the British Isles.* London: Allen and Unwin.

Pethick, J., 1993, Shoreline adjustments and coastal management: physical and biological processes under accelerated sea level rise. *Geographical Journal* 159, 162–8.

Petts, G. E., 1985, *Impounded Rivers: Perspectives for Ecological Management.* Chichester: Wiley.

Petts, G. E., 1988, Water management: the case of Lake Biwa, Japan. *Geographical Journal* 154, 367–76.

Pickering, K. T. and Owen, L. A., 1994, *An Introduction to Global Environmental Issues.* London: Routledge.

Pimental, D. (ed.), 1993, *World Soil Erosion and Conservation.* Cambridge: Cambridge University Press.

Piotrowska, H., 1989, Natural and anthropogenic changes in sand-dunes and their vegetation on the southern Baltic coast. In F. van der Meulen, P. D. Jungerius, and J. Visser (eds), *Perspectives in Coastal Dune Management,* 33–40. The Hague: SPB Academic Publishing.

Ponting, C., 1991, *A Green History of the World.* London: Penguin.

Primavera, J. H., 1995, Mangroves and brackish water pond culture in the Philippines. *Hydrobiologia* 295, 303–9.

Pye, K., 1990, Physical and human influences on coastal dune development between the Ribble and the Mersey estuaries, NW England. In K. F. Nordstrom, N. P. Psuty and R. W. G. Carter (eds), *Coastal Dunes,* 339–59. Chichester: Wiley.

Pyne, S. J., 1982, *Fire in America: A Cultural History of Wildland and Rural Fire.* Princeton: Princeton University Press.

Ramphele, M., 1991, *Restoring the Land: Environment and Change in Post-Apartheid South Africa.* London: Panos.

Ranwell, D. S. and Boar, R., 1986, *Coast Dune Management Guide.* Huntingdon: Institute of Terrestrial Ecology.

Rhoades, J. D., 1990, Soil salinity: causes and controls. In A. S. Goudie (ed.), *Techniques for Desert Reclamation*, 109–34. Chichester: Wiley.

Ripley, E. A., 1976, Drought in the Sahara: insufficient geophysical feedback? *Science* 191, 100.

Robertson, D. G. and Slack, R. D., 1995, Landscape change and its effects on the wintering range of a Lesser Snow Goose *Chen caerulescens caerulescens* population: a review. *Biological Conservation* 71, 179–85.

Robinson, M., 1990, *Impact of improved land drainage on river flows*. Institute of Hydrology, Wallingford, report no. 113.

Romme, W. H. and Despain, D. G., 1989, The Yellowstone fires. *Scientific American* 261, 21–9.

Rozanov, B. G., Targulian, V. and Orlov, D. S., 1990, Soils. In B. L. Turner, W. C. Clark, R. W. Kates, J. F. Richards, J. T. Matthews and W. B. Meyer (eds), *The Earth as Transformed by Human Action*, 203–14. Cambridge: Cambridge University Press.

Sahagian, D. L., Schwartz, F. W. and Jacobs, D. K., 1994, Direct anthropogenic contributions to sea level rise in the twentieth century. *Nature* 367, 54–7.

Schmid, J. A., 1975, *Urban Vegetation*. University of Chicago, Geography Department research paper no. 161.

Schneider, S. H. and Thompson, S. L., 1988, Simulating the effects of nuclear war. *Nature* 333, 221–7.

Schneider, W. J., 1970, *Hydrological Implications of Solid-waste Disposal*. United States Geological Survey Circular no. 601-F.

Schulze, E.-D., Lange, O. L. and Oren, R., 1989, *Forest Decline and Air Pollution*. Ecological Studies no. 71. New York: Springer-Verlag.

Schwarz, H. E., Emel, J., Dickens, W. J.,

Rogers, P. and Thompson, J., 1990, Water quality and flows. In B. L. Turner, W. C. Clark, R. W. Kates, J. F. Richards, J. T. Matthews and W. B. Meyer (eds), *The Earth as Transformed by Human Action*, 253–70. Cambridge: Cambridge University Press.

Shiklomanov, I. A., 1985, Large scale water transfers. In J. C. Rodda (ed.), *Facets of Hydrology II*, 345–87. Chichester: Wiley.

Shukla, J., Nobre, C. and Sellers, P., 1990, Amazon deforestation and climatic change. *Science* 247, 1322–5.

Simmons, I. G., 1989, *Changing the Face of the Earth: Culture, Environment and History*. Oxford: Blackwell.

Simmons, I., 1993, *Environmental History: A Concise Introduction*. Oxford: Blackwell.

Smith, J. B., Carmichael, J. J. and Titus, J. G., 1995, Adaptation policy. In K. M. Strzepek and J. B. Smith (eds), *As Climate Changes: International Impacts and Implications*, 201–10. Cambridge: Cambridge University Press.

Spate, O. H. K. and Learmonth, A. T. A., 1967, *India and Pakistan*. London: Methuen.

Swanston, D. N. and Swanson, F. J., 1976, Timber harvesting, mass erosion and steepland forest geomorphology in the Pacific Northwest. In D. R. Coates (ed.), *Geomorphology and Engineering*, 199–221. Stroudsberg, PA: Dowden, Hutchinson and Ross.

Thomas, D. S. G. and Middleton, N. J., 1993, Salinization: new perspectives on a major issue. *Journal of Arid Environments* 24, 95–105.

Thomas, D. S. G. and Middleton, N. J., 1994, *Desertification: Exploding the Myth*. Chichester: Wiley.

Tickell, C., 1993, The human species: a suicidal success? *Geographical Journal* 159, 215–26.

Tiffen, M., Mortimore, M. and Gichuki,

F., 1994, *More People, Less Erosion: Environmental Recovery in Kenya*. Chichester: Wiley.

Tolba, M. K. and El-Kholy, O. A. (eds), 1992, *The World Environment, 1972–1992*. London: UNEP/Chapman and Hall.

Trimble, S. W., 1974, *Man-induced Soil Erosion on the Southern Piedmont*. Ankeny, Iowa: Soil Conservation Society of America.

Turco, R. P., Toon, O. B., Ackermann, T. P., Pollack, J. B. and Sagan, C., 1983, Nuclear winter: global consequences of multiple nuclear explosions. *Science* 222, 1283–92.

Turner, B. L., Clark, W. C., Kates, R. W., Richards, J. F., Matthews, J. T. and Meyer, W. B. (eds), 1990, *The Earth as Transformed by Human Action*. Cambridge: Cambridge University Press.

Turner, B. L., Kasperson, R. E., Meyer, W. B., Dow, K. M., Golding, D., Kasperson, J. X., Mitchell, R. C. and Ratick, S. J., 1990, Two types of global environmental change: definitional and spatial-scale issues in their human dimensions. *Global Environmental Change* 1, 14–22.

UNEP, 1989, *Environmental Data Report 1989–90*. Oxford: Blackwell/United Nations Environmental Programme.

UNEP, 1993, *Environmental Data Report 1993–4*. Oxford: Blackwell/United Nations Environmental Programme.

Usoro, E. J., 1985, Nigeria. In E. C. F. Bird and M. L. Schwartz (eds), *The World's Coastline*, 607–13. New York: Van Nostrand Reinhold.

Viles, H. A. and Spencer, T., 1995, *Coastal Problems*. London: Edward Arnold.

Vitousek, P. M., 1994, Beyond global warming: ecology and global change. *Ecology* 75, 1861–76.

Vogel, C. H. and Drummond, J. H., 1995, Shades of 'green' and 'brown': environmental issues in South Africa. In A. Lemon (ed.), *The Geography of Change in South Africa*, 85–98. Chichester: Wiley.

Wahren, C.-H., Papst, W. A. and Williams, R. J., 1994, Long-term vegetation change in relation to cattle grazing in subalpine grassland and heathland in the Bugong High Plains: an analysis of vegetation records from 1945 to 1994. *Australian Journal of Botany* 42, 607–39.

Walsh, R. P., Hudson, R. N. and Howells, K. A., 1982, Changes in the magnitude-frequency of flooding and heavy rainfalls in the Swansea valley since 1875. *Cambria* 9(2), 36–60.

Waltham, A. C., 1991, *Land Subsidence*. Glasgow: Blackie.

Ward, R. C., 1978, *Floods: A Geographical Perspective*. London: Macmillan.

Warrick, R. A. and Oerlemans, J., 1990, Sea level rise. In J. T. Houghton, G. J. Jenkins and J. J. Ephraums, *Climate Change: The IPCC Scientific Assessment*, 257–81. Cambridge: Cambridge University Press.

Watson, R. T., Zinyowera, M. C. and Moss, R. H., 1996, *Climate Change 1995 – Impacts, Adaptations and Mitigation of Climate Change: Scientific–Technical Analyses*. Contribution of Working Group II to the Second Assessment Report of the Intergovernmental Panel on Climate Change. Cambridge: Cambridge University Press.

Weber, P., 1993, Reviving coral reefs. In L. R. Brown (ed.), *State of the World 1993*, 42–60. London: Earthscan.

Wellburn, A., 1988, *Air Pollution and Acid Rain: The Biological Impact*. Harlow: Longman.

Wells, S. M., 1988, *Coral Reefs of the World*, vol. 2: *Indian Ocean, Red Sea and Gulf*. Gland, Switzerland/Cambridge: UNEP/IUCN (International Union for the Conservation of Nature).

White, R., 1994, *Urban Environmental Management*. Chichester: Wiley.

Whitmore, T. M., Turner, B. L., Johnson, D. L., Kates, R. W. and Gottschang, T. R., 1990, Long term population change. In B. L. Turner, W. C. Clark, R. W. Kates, J. F. Richards, J. T. Matthews and W. B. Meyer (eds), *The Earth as Transformed by Human Action*, 26–39. Cambridge: Cambridge University Press.

Wigley, T. M. L. and Raper, S. C. B., 1992, Implications for climate and sea level of revised IPCC emissions scenarios. *Nature* 357, 293–300.

Wigley, T. M. L. and Raper, S. C. B., 1993, Future changes in global mean temperatures and sea level. In R. A. Warrick, E. M. Barrow and T. M. L. Wigley (eds), *Climate and Sea Level Change*, 111–33. Cambridge: Cambridge University Press.

Wilcove, D. S., McLellan, C. H. and Dobson, A. P., 1986, Habitat fragmentation in the temperate zone. In M. E. Soulé (ed.), *Conservation Biology: The Science of Scarcity and Diversity*, 251–6. Sunderland, Mass.: Sinauer Associates.

Williams, M., 1989, *Americans and their Forests*. Cambridge: Cambridge University Press.

Williams, M., 1994, Forests and tree cover. In W. B. Meyer and B. L. Turner (eds), *Changes in Land Use and Land Cover: A Global Perspective*, 97–124. Cambridge: Cambridge University Press.

Williams, M. (ed.), 1990, *Wetlands: A Threatened Landscape*. Oxford: Blackwell.

Wilson, E. O., 1992, *The Diversity of Life*. London: Penguin.

Winkler, E. M., 1975, *Stone: Properties, Durability in Man's Environment*. Vienna: Springer-Verlag.

Woodcock, N., 1994, *Geology and Environment in Britain and Ireland*. London: University College London Press.

World Resources Institute, 1994, *World Resources 1994/5*. New York: Oxford University Press.

Worthington, E. B. (ed.), 1977, *Arid Land Irrigation in Developing Countries: Environmental Problems and Effects*. Oxford: Pergamon.

Yim, W. W.-S., 1993, Future sea level rise in Hong Kong and possible environmental effects. In R. A. Warrick, E. M. Barrow and T. M. L. Wigley (eds), *Climate and Sea Level Change*, 349–76. Cambridge: Cambridge University Press.

INDEX

Note: Alphabetical arrangement of headings and subheadings is word by word, ignoring 'and', 'by', 'in', 'through', etc. Page numbers in *italics* refer to illustrations. Remember to consult the Glossary (pp. 245–53) for definitions.

Index compiled by Ann Barham